# Induction of Pine Volatiles by Insect Egg Deposition

**Inaugural-Dissertation**
**zur Erlangung des Doktorgrades**
**am Fachbereich Biologie, Chemie, Pharmazie der**
**Freien Universität Berlin**

**angefertigt am Institut für Biologie**
**Angewandte Zoologie / Ökologie der Tiere**

von

**Roland Mumm**
**aus Heide**

Berlin, im August 2004

cover design by: Nina Fatouros
photos by: Cornelia Dippel, Kai Schrank

Bibliographic information published by Die Deutsche Bibliothek

Die Deutsche Bibliothek lists this publication in the Deutsche Nationalbibliografie;
detailed bibliographic data is available in the Internet at http://dnb.ddb.de.

ISBN 3-8325-0846-5

Logos Verlag Berlin
Comeniushof, Gubener Str. 47,
10243 Berlin
Tel.: +49 030 42 85 10 90
Fax: +49 030 42 85 10 92
INTERNET: http://www.logos-verlag.de

1. Gutacher: Prof. Dr. Monika Hilker

2. Gutachter: Prof. Dr. Johannes Steidle

*Meinen Eltern, Daniel & Sylvia*

*This thesis is based on the following manuscripts:*

1) Monika Hilker, Claudia Stein, Roland Schröder, Martii Varama, and Roland Mumm (submitted). Insect egg deposition induces defence responses in *Pinus sylvestris*: Characterisation of the elicitor. (A revised version of the manuscript is accepted by *Journal of Experimental Biology*)

2) Roland Mumm, Kai Schrank, Robert Wegener, Stefan Schulz, and Monika Hilker (2003). Chemical analysis of volatiles emitted by *Pinus sylvestris* after induction by insect oviposition. *Journal of Chemical Ecology* 29: 1235-1252.

3) Roland Mumm and Monika Hilker. How does an egg parasitoid recognize odour of pine with host eggs ? (A revised version of the manuscript is accepted by *Chemical Senses*)

4) Roland Mumm, Tassilo Tiemann, Martti Varama, and Monika Hilker (submitted). Choosy egg parasitoids: Specificity of oviposition-induced pine volatiles exploited by an egg parasitoid of pine sawflies. (A revised version of the manuscript is accepted by *Entomologia Experimentalis et Applicata*)

5) Roland Mumm, Tassilo Tiemann, Stefan Schulz, and Monika Hilker (submitted). Analysis of volatiles from black pine (*Pinus nigra*): Significance of wounding and egg deposition by a herbivorous sawfly. (A revised version of the manuscript is published in *Phytochemistry* 65: 3221-3230, 2004).

6) Roland Mumm and Monika Hilker (manuscript). Defence in pines (*Pinus* spp.) to biotic stress by herbivorous insects.

# Contents

# Chapter 1

# General Introduction and Thesis Outline

Plants as autotrophic primary producers represent the first trophic level and serve the essential basis for all heterotrophic organisms. They are threatened by a myriad of potential attackers as herbivorous arthropods, nematodes, pathogenic microorganisms, and vertebrates. Since plants can not run away or hide from attacks, a great variety of defensive mechanisms against the different threats have evolved.

Plant defence mechanisms may be constitutively present irrespective whether there is an actual need to defend against. Constitutive (preformed) defences may form a first physical and/or chemical barrier against attack. Such defences comprise morphological traits as thorns that may deter herbivores from feeding or trichomes which can hamper the movement of small insects or also can kill them. In addition, plants continually produce and accumulate a plethora of secondary metabolites (e.g. terpenoids, phenols, tannins) that may affect the behaviour or physiology of herbivores (overview given by Rosenthal and Berenbaum 1991). However, maintenance of constitutive defences produce physiological costs (nutrients and energy) which can not be allocated for growth and reproduction (Herms and Mattson 1992; Karban and Baldwin 1997).

In contrast, inducible defences are expressed only after an initial attack and may allow plants to minimize fitness costs of resistance (Cipollini et al. 2003; Heil and Baldwin 2002; Karban and Baldwin 1997). Inducible defences are considered as a form of phenotypic plasticity allowing plants to change their phenotype in response to damage caused by herbivores or changing environmental conditions (Agrawal 2001; Dicke and Hilker 2003). Additionally, inducibility may reduce the chance that particularly specialised herbivores exhaust these defensive traits, e.g. for their own defence or chemical communication mechanisms, thus decreasing so called ecological costs (Agrawal and Karban 1999; Cipollini et al. 2003; Zangerl 2003, and references therein). Moreover, the release of plant volatiles only when under herbivore attack provides a reliable and "honest" signal for potential natural enemies (Zangerl 2003; see below).

Inducible defences can be classified as direct defences that affect the herbivore's biology directly and indirect defences which affect the herbivore by recruiting its natural enemies (Dicke and Vet 1999; Dicke et al. 2003). Indirect defence responses induced by herbivorous insects comprise the production of extrafloral nectar that is exploited by carnivores (Heil et al. 2001; Wäckers et al. 2001) as well as the induced production of plant volatiles that attract carnivorous arthropods like predators and parasitoids (Dicke and van Loon 2000; Sabelis et al. 1999; Turlings et al. 2002). Thus, direct and indirect induced plant defences can influence the bottom-up and top-down forces affecting herbivore population dynamics (Dicke and Hilker 2003).

It is well known that parasitoids use chemical cues when actively foraging for hosts (e.g. Vinson 1998). These so called infochemicals (Dicke and Sabelis 1988) can originate either directly from the herbivore or the host plant where the herbivore feeds on (Vet and Dicke 1992). Parasitoids are faced with what has been termed "reliability-detectebility-dilemma", i.e. cues from the herbivores may be highly reliable, but are less detectable compared to volatiles of plants which have a much larger biomass (Dicke 1999a; Vet and Dicke 1992). Plant volatiles that are induced by feeding herbivores may therefore be both detectable and reliable indicators of herbivores (e.g. Dicke 1999b; Turlings et al. 2002). Plants generally emit larger amounts of volatile constituents after herbivore feeding compared to uninfested plants (Paré and Tumlinson 1999). This is partly due to stored compounds in the plant tissue that are released upon the mechanical damage. Moreover, the volatile blend emitted by plants after herbivore feeding can be qualitatively and/or quantitatively different from volatiles emitted from artificially wounded ones (Paré and Tumlinson 1999; Turlings et al. 1998). Herbivore-induced plant volatiles can be very specific, thus providing carnivores with reliable information about the plant species (Geervliet et al. 1997; Gouinguené et al. 2001; Takabayashi and Dicke 1996), herbivore species (De Moraes et al. 1998; Du et al. 1998; Turlings et al. 1998, 2002), and even developmental stage (Gouinguené et al. 2003; Takabayashi et al. 1995).

Another possibility how parasitoids can link highly detectable to highly reliable cues is through associative learning (Vet et al. 1995). The ability to learn odours is assumed to be especially adaptive in generalist parasitoids that have a broad host range which also feed on many host plants and therefore have to deal with a great odour variability (Steidle and van Loon 2003; Vet and Dicke 1992). Conversely, specialised parasitoids primarily show innate responses to specific and reliable cues (Steidle and van Loon 2003).

In contrast to constitutive defences, inducible defences need additional mechanisms to perceive (recognise) the presence of herbivores or pathogens accurately. After perception of the stimulus, the plant transmits the signal inside the cell across the plasma membrane and also to other cells resulting into the induction of defence responses (signal transduction). Many signal transduction pathways are mediated by plant hormones which play an important role in signalling defence responses against herbivores (de Bruxelles and Roberts 2001). For example, jasmonic acid (JA) which is a product of the lipoxygenase pathway, is a central molecule in wound-induced defence responses (Staswick and Lehman 1999). It has been demonstrated that the exogenous application of JA or its volatile derivative methyl jasmonate (MeJA) induces direct and indirect defence responses (e.g. Baldwin 1998; Gols et al. 1999; Ozawa et al. 2000).

Mechanical wounding usually can not mimic the indirect defence responses induced by herbivores as parasitoids can discriminate between these volatile blends (reviewed by Turlings et al. 2002). This can be explained by the presence of herbivore borne elicitors. For example, the application of oral secretions of herbivore caterpillars into artificially wounded plant tissue can induce similar responses as herbivory itself (Tumlinson et al. 1999; Turlings et al. 1990). Two types of active elicitors have been identified from the oral secretions of herbivores: (1) a β-glucosidase has been isolated from the regurgitate of *Pieris brassicae* larvae (Mattiacci et al. 1995); (2) fatty acid – amino acid conjugates, e.g. volicitin have been isolated from the regurgitate of several lepidopteran species (Alborn et al. 1997, 2000; Halitschke et al. 2001; Pohnert et al. 1999). Furthermore, a glucose oxidase that was isolated from the saliva of *Heliothis zea* elicits a suppression of the plant's defensive response (Musser et al. 2002).

The induction of defence responses may not be locally restricted to the site of attack but systemic responses have been shown, i.e. distant undamaged parts of the plant are induced to produce defensive compounds or to emit volatiles attractive to carnivores (Dicke et al. 1993; McAuslane and Alborn 1998; Röse et al. 1998). Systemic responses require mobile signals that are transported by the vasculature from the infested sites to other plant parts. Several chemical signals as well as electrical and hydraulic signals have been proposed as systemic wound signals (reviewed by de Bruxelles and Roberts 2001).

Despite the advantages of inducible defences against feeding herbivores, one obvious drawback is the time-lag until the defence becomes active. During this time the plant is still vulnerable to herbivory (Zangerl 2003). One way to prevent extensive feeding damage by larvae may be to defend already against the egg stage. Indeed, several studies demonstrated that the egg deposition of herbivore females can already induce direct and indirect defence responses in plants (reviewed by Hilker and Meiners 2002, Hilker et al. 2002b). Egg deposition of herbivores can induce the release of plant volatiles that attract specialised egg parasitoids (Colazza et al. 2004, Hilker et al. 2002a; Meiners and Hilker 2000). This plant response has been interpreted as a preventive defensive strategy against herbivory since plants react to the herbivore attack prior to being damaged by the next herbivore generation (Hilker and Meiners 2002).

In the tritrophic system investigated here the egg deposition of the herbivorous conifer sawfly *Diprion pini* L. (Hymenoptera, Diprionidae) is known to induce Scots pine (*Pinus sylvestris* L.) to release volatiles that attract experienced females of *Chrysonotomyia ruforum* Krausse (Hymenoptera, Eulophidae), an egg parasitoid specialised on eggs of diprionid sawflies (overview given in Tab. 1; Hilker et al. 2002a). Oviposition by *D. pini* induces not only a local response, but also a systemic reaction in pine needles without eggs but adjacent to those carrying sawfly eggs. Pine twigs treated with jasmonic acid also emit volatiles that attract the egg parasitoids. Volatiles released from pine twigs without eggs that were artificially wounded to mimic the mechanical damage inflicted by the ovipositor of the female sawfly are not attractive. Applying the oviduct secretion that coats the eggs into wounded pine needles induces volatiles attractive to the egg parasitoids. Thus, the elicitor responsible for the volatile induction is located in the oviduct secretion of *D. pini* females.

**Table 1** Response of the experienced egg parasitoid *C. ruforum* to differently treated twigs of *P. sylvestris* as published by Hilker et al. (2002a). + indicate that parasitoids were significantly ($p<0.05$) attracted to pine volatiles in the olfactometer, 0 indicates no significant ($p>0.05$) responses to the offered pine odours (Friedman ANOVA, Wilcoxon-Wilcox-test for multiple comparisons). For details see text.

| Volatiles of | Scale | Response of experienced *C. ruforum* females |
|---|---|---|
| pine twigs with eggs of *D. pini* | local / systemic | + |
| pine twigs without eggs of *D. pini* | local / systemic | 0 |
| artificially wounded pine twigs | local | 0 |
| artificially wounded pine twigs + freshly obtained oviduct secretion | local | + |
| pine twigs treated with jasmonic acid (JA) | local | + |

Based on these results the main goal of this thesis is to provide further information on the understanding of the mechanisms underlying these tritrophic interactions between pine, the sawfly, and the egg parasitoid.

Experiments presented in *chapter 2* aimed to further characterize the physical and chemical properties of the elicitor in the oviduct secretion of *D. pini* females which is responsible for the volatile induction in pine.

In *chapter 3*, results of chemical analyses are presented to elucidate the pine volatile pattern induced by egg deposition of *D. pini*. The question was addressed whether egg deposition or treatment with jasmonic acid induce qualitative or quantitative changes in the volatile blend of pine. Detailed chemical analyses of the headspace volatiles of oviposition- and JA-induced pine twigs were made and compared with volatiles released of respective controls.

Based on the differences found between the headspace volatiles of induced and non-induced pine, olfactometer bioassays were conducted to test whether single compounds out of the volatile blend are important for the host orientation of the egg parasitoid *C. ruforum*. These assays are described in *chapter 4*.

*Chapter 5* deals with the specificity of chemical cues used by the egg parasitoid *C. ruforum* for host search. The response of naïve female egg parasitoids, which had no previous experience with a plant-host complex, to oviposition-induced pine volatiles was investigated. Further it was investigated whether response of the egg parasitoids to pine volatiles after egg deposition is specific for the pine species or the sawfly species. In addition, it was studied whether larval feeding also induces volatiles in *P. sylvestris* that attract female *C. ruforum*.

In *chapter 6*, results of the chemical analyses of the headspace volatiles of black pine (*Pinus nigra*) are presented. Volatiles of this pine species carrying eggs of *D. pini* were not attractive to the egg parasitoids (see *chapter 5*). Therefore, the headspace of untreated, egg-carrying, and artificially wounded *P. nigra* twigs were analysed by coupled gas chromatography – mass spectrometry (GC-MS). and compared by multivariate statistics with the volatile composition of *P. sylvestris*.

*Chapter 7* reviews the different constitutive and induced defence mechanisms of pines against herbivores embedding the information gained in this thesis and setting them into a broader ecological context.

# References

Agrawal, AA (2001). Phenotypic plasticity in the interactions and evolution of species. Science 294:321-326.

Agrawal, AA, Karban, R (1999). Why induced defenses may be favored over constitutive strategies in plants. In: Tollrian, R, Harvell, CD (eds), The Ecology and Evolution of Inducible Defenses. Princeton University Press, pp. 45-61.

Alborn, HT, Jones, TH, Stenhagen, G, Tumlinson, JH (2000). Identification and synthesis of volicitin and related components from beet armyworm oral secretions. J. Chem. Ecol. 26:203-220.

Alborn, HT, Turlings, TCJ, Jones, TH, Stenhagen, G, Loughrin, JH, Tumlinson, JH (1997). An elicitor of plant volatiles from beet armyworm oral secretion. Science 276:945-949.

Baldwin, IT (1998). Jasmonate-induced responses are costly but benefit plants under attack in native populations. Proc. Natl. Acad. Sci. USA 95:8113-8118.

Cipollini, D, Purrington, CB, Bergelson, J (2003). Costs of induced responses in plants. Basic Appl. Ecol. 4 :79-85.

Colazza, S, Fucarino, A, Peri, E, Salerno, G, Conti, E, Bin, F (2004). Insect oviposition induces volatile emission in herbaceous plants that attracts egg parasitoids. J. Exp. Biol. 207:47-53.

de Bruxelles, GL, Roberts, MR (2001). Signals regulating multiple responses to wounding and herbivores. Crit. Rev. Plant Sci. 20:487-521.

De Moraes, CM, Lewis, WJ, Paré, PW, Alborn, HT, Tumlinson, JH (1998). Herbivore-infested plants selectively attract parasitoids. Nature 393:570-573.

Dicke, M (1999a). Evolution of induced indirect defense of plants. In: Tollrian, R, Harvell, CD (eds), The Ecology and Evolution of Inducible Defenses. Princeton University Press, Princeton, pp. 62-88.

Dicke, M (1999b). Are herbivore-induced plant volatiles reliable indicators of herbivore identity to foraging carnivorous arthropods? Entomol. Exp. Appl. 91:131-142.

Dicke, M, Hilker, M (2003). Induced plant defences: from molecular biology to evolutionary ecology. Basic Appl. Ecol. 4:3-14.

Dicke, M, van Loon, JJA (2000). Multitrophic effects of herbivore-induced plant volatiles in an evolutionary context. Entomol. Exp. Appl. 37:237-249.

Dicke, M, Vet, LEM (1999). Plant-carnivore interactions: evolutionary and ecological consequences for plant, herbivore and carnivore. In: Olff, H, Brown, VK, Drent, RH (eds), Herbivores: Between Plants and Predators. Blackwell Science, pp. 483-520.

Dicke, M, Sabelis, MW (1988). Infochemical terminology: based on cost-benefit analysis rather than origin of compounds? Funct. Ecol. 2:131-139.

Dicke, M, van Poecke, RMP, de Boer, JG (2003). Inducible indirect defence of plants: from mechanisms to ecological functions. Basic Appl. Ecol. 4:27-42.

Dicke, M, Van Baarlen, P, Wessels, R, Dijkman, H (1993). Herbivory induces systemic production of plant volatiles that attract predators of the herbivore: extraction of endogenous elicitor. J. Chem. Ecol. 19:581-599.

Du, Y, Poppy, GM, Powell, W, Pickett, JA, Wadhams, LJ, Woodcock, CM (1998). Identification of semiochemicals released during feeding that attract Parasitoid Aphidius ervi. J. Chem. Ecol. 24:1355-1368.

Geervliet, JBF, Posthumus, MA, Vet, LEM, Dicke, M (1997). Comparative analysis of headspace volatiles from different caterpillar-infested or uninfested food plants of Pieris species. J. Chem. Ecol. 23:2935-2954.

Gols, R, Posthumus, MA, Dicke, M (1999). Jasmonic acid induces the production of gerbera volatiles that attract the biological control agent Phytoseiulus persimilis. Entomol. Exp. Appl. 93:77-86.

Gouinguené, S, Alborn, HT, Turlings, TCJ (2003). Induction of volatile emissions in maize by different larval instars of *Spodoptera littoralis*. J. Chem. Ecol. 29:145-162.

Gouinguené, S, Degen, T, Turlings, TCJ (2001). Variability in herbivore-induced odour emissions among maize cultivars and their wild ancestors (teosinte). Chemoecology 11:9-16.

Halitschke, R, Schittko, U, Pohnert, G, Boland, W, Baldwin, IT (2001). Molecular interactions between the specialist herbivore *Manduca sexta* (Lepidoptera, Sphingidae) and its natural host N*icotiana attenuata*. III. Fatty acid-amino acid conjugates in herbivore oral secretions are necessary and sufficient for herbivore-specific plant responses. Plant Physiol. 125:711-717.

Heil, M, Baldwin, IT (2002). Fitness costs of induced resistance: emerging experimental support for a slippery concept. Trends Plant Sci. 7:61-67.

Heil, M, Koch, T, Hiplert, A, Fiala, B, Boland, W, Linsenmair, KE (2001). Extrafloral nectar production of ant-associated plant *Macaranga tanarius*, is an induced, indirect defensive response elicited by jasmonic acid. Proc. Natl. Acad. Sci. USA 98:1083-1088.

Herms, DA, Mattson, WJ (1992). The dilemma of plants: to grow or defend. Quart. Rev. Biol. 67:283-335.

Hilker, M, Meiners, T (2002). Induction of plant responses towards oviposition and feeding of herbivorous arthropods: a comparison. Entomol. Exp. Appl. 104:181-192.

Hilker, M, Kobs, C, Varama, M, Schrank, K (2002a). Insect egg deposition induces *Pinus* to attract egg parasitoids. J. Exp. Biol. 205:455-461.

Hilker, H, Rohfritsch, O, Meiners, T (2002b). The plant's response towards insect egg deposition. In: Hilker, M, Meiners, T (eds), Chemoecology of Insect Eggs and Egg Deposition. Blackwell Publishing, Berlin, Oxford, pp. 205-233.

Karban, R, Baldwin, IT (1997). Induced responses to herbivory. Chicago University Press, Chicago.

Mattiacci, L, Dicke, M, Posthumus, MA (1995). β-Glucosidase: An elicitor of herbivore-induced plant odor that attracts host-searching parasitic wasps. Proc. Natl. Acad. Sci. USA 92:2036-2040.

McAuslane, HJ, Alborn, HT (1998). Systemic induction of allelochemicals in glanded and glandless isogenic cotton by *Spodoptera exigua* feeding. J. Chem. Ecol. 24:399-416.

Meiners, T, Hilker, M (2000). Induction of plant synomones by oviposition of a phytophagous insect. J. Chem. Ecol. 26:221-232.

Musser, RO, Hum-Musser, SM, Eichenseer, H, Pfeiffer, M, Ervin, G, Murphy, JB, Felton, GW (2002). Caterpillar saliva beats plant defences. Nature 116:599-600.

Ozawa, R, Arimura, G, Takabayashi, J, Shimoda, T, Nishioka, T (2000). Involvement of jasmonate- and salicylate-related signaling pathways for the production of specific herbivore-induced volatiles in plants. Plant Cell Physiol. 41:391-398.

Paré, PW, Tumlinson, JH (1999). Plant volatiles as a defense against insect herbivores. Plant Physiol. 121:325-331.

Pohnert, G, Jung, V, Haukioja, E, Lempa, K, Boland, W (1999). New fatty acid amides from regurgitant of lepidopteran (Noctuidae, Geometridae) caterpillars. Tetrahedron 55:11275-11280.

Rosenthal, GA, Berenbaum, MR (1991). Herbivores - Their Interactions with Secondary Metabolites Vol. 1 The Chemical Participants. Academic Press, San Diego.

Röse, USR, Lewis, WJ, Tumlinson, JH (1998). Specificity of systemically released cotton volatiles as attractants for specialist and generalist parasitic wasps. J. Chem. Ecol. 24:303-319.

Sabelis, M, Janssen, A, Pallini, A, Venzon, M, Bruin, J, Drukker, B, Scutareanu, P (1999). Behavioral responses of predatory and herbivorous arthropods to induced plant volatiles: from evolutionary ecology to agricultural applications. In: Agrawal, AA, Tuzun, S, Bent, E (eds), Induced Plant Defenses against Pathogens and Herbivores. APS Press, St. Paul, pp. 269-296.

Staswick, PE, Lehman, CC (1999). Jasmonic acid-signalled responses in plants. In: Agrawal, AA, Tuzun, S, Bent, E (eds), Induced Plant Defenses against Pathogens and Herbivores. APS Press, St. Paul, Minnesota, pp. 117-136.

Steidle, JLM, van Loon, JJA (2003). Dietary specialization and infochemical use in carnivorous arthropods: testing a concept. Entomol. Exp. Appl. 108:133-148.

Takabayashi, J, Dicke, M (1996). Plant-carnivore mutualism through herbivore-induced carnivore attractants. Trends Plant Sci. 1:109-113.

Takabayashi, J, Takahashi, S, Dicke, M, Posthumus, M.A. (1995). Developmental stage of herbivore *Pseudaletia separata* affects production of herbivore -induced synomone by corn plants. J. Chem. Ecol. 21:273-287.

Tumlinson, JH, Paré, P, Lewis, WJ (1999). Plant production of volatile semiochemicals in response to insect-derived elicitors. In: Chadwick, DJ, Goode, JA (eds), Insect-Plant Interactions and Induced Plant Defence. Wiley & Sons, Chichester, pp. 95-105.

Turlings, TCJ, Gouinguené, S, Degen, T, Fritzsche-Hoballah, ME (2002). The chemical ecology of plant-caterpillar-parasitoid interactions. In: Tscharntke, T, Hawkins, B (eds), Multitrophic Level Interactions. Cambridge University Press, Cambridge, pp. 148-173.

Turlings, TCJ, Bernasconi, M, Bertossa, R, Bigler, F, Caloz, G, Dorn, S (1998). The induction of volatile emissions in maize by three herbivore species with different feeding habitats: possible consequences for their natural enemies. Biol. Control 11:122-129.

Turlings, TCJ, Tumlinson, JH, Lewis, WJ (1990). Exploitation of herbivore-induced plant odors by host-seeking parasitic wasps. Science 250:1251-1253.

Vet, LEM, Dicke, M (1992). Ecology of infochemical use by natural enemies in a tritrophic context. Annu. Rev. Entomol. 37:141-172.

Vet, LEM, Lewis, WJ, Cardé, RT (1995). Parasitoid foraging and learning. In: Cardé, RT, Bell, WJ (eds), Chemical Ecology of Insects 2. Chapman and Hall, London; New York, pp. 65-101.

Vinson, SB (1998). The general host selection behavior of parasitoid hymenoptera and a comparison of initial strategies utilized by larvaphagous and oophagous species. Biol. Control 11:79-96.

Wäckers, FL, Zuber, D, Wunderlin, R, Keller, F (2001). The effect of herbivory on temporal and spatial dynamics of foliar nectar production in cotton and castor. Annals Bot. 87:365-370.

Zangerl, AR (2003). Evolution of induced plant responses to herbivores. Basic Appl. Ecol. 4:91-103.

# Chapter 2

# Insect Egg Deposition Induces Defence Responses in
# *Pinus sylvestris*: Characterisation of the Elicitor

**Abstract.** Egg deposition by the phytophagous sawfly *Diprion pini* L. (Hymenoptera, Diprionidae) is known to locally and systemically induce the emission of volatiles in Scots pine (*Pinus sylvestris* L.) that attract the egg parasitoid *Chrysonotomyia ruforum* Krausse (Hymenoptera, Eulophidae). The egg parasitoids are killing the eggs, thus preventing the plant from feeding damage by sawfly larvae. The elicitor inducing the pine's response is known to be located in the oviduct secretion which the female sawfly applies to the eggs when inserting them into a pine needle slit by the sclerotized ovipositor valves. In this study, the elicitor located in the oviduct secretion was characterized. The elicitor kept its inducing activity when applied directly after isolation from the oviduct into artificially slit pine needles. However, as soon as the oviduct secretion was solved in Aqua dest. and stored for 3 h at room temperature or kept frozen at –80 °C , its activity was lost. In contrast, oviduct secretion kept its eliciting activity, when solved in Ringer solution (pH 7.2) both after storage at room temperature and freezer storage. The activity of the elicitor vanished after treatment of the oviduct secretion with proteinase K, which destroyed all proteins. This suggests that the elicitor in the oviduct secretion is a peptide or protein, or a component bound to these. An SDS-PAGE revealed a similar, but not the same protein pattern of hemolymph and oviduct secretion. Hemolymph itself has no eliciting effect. The elicitor in the oviduct secretion is only active when transferred on slit pine needles, since its application on undamaged needles did not induce the emission of attractive volatiles.

**Keywords.** egg deposition, egg parasitoid, elicitor, induction, oviduct secretion, sawfly, volatiles

**Introduction**

Both insect egg deposition and feeding by herbivorous arthropods is well-known to induce plant defensive responses. The plant's response may have direct detrimental effects on the eggs, the ovipositing female or the feeding herbivore (Hilker et al. 2002b). Furthermore, the plant is able to respond to egg deposition and feeding by recruiting predators and parasitoids of the herbivores (Baldwin and Preston 1999, Dicke and van Loon 2000; Hilker and Meiners 2002; Turlings et al. 2002). Especially plant volatiles induced by egg deposition or feeding have been shown to attract enemies of the herbivores (Boland et al. 1999; Colazza et al. 2004a,b; Dicke and Hilker 2003; Hilker and Meiners 2002; Mumm et al. 2003). While the plant responding to feeding herbivores acts "online" with the damage, the plant responding to insect egg deposition acts prior to being damaged by feeding larvae. Thus, Hilker and Meiners (2002) interpreted the plant's defensive response to insect egg deposition as a preventive defence mechanism.

Egg deposition by the phytophagous sawfly *Diprion pini* L. (Hymenoptera, Diprionidae) has been shown to induce the plant to release volatiles attracting *Chrysonotomyia ruforum* Krausse (Hymenoptera, Eulophidae) an egg parasitoid of *D. pini* (Hilker et al. 2002a). This induction of volatiles is not restricted to the oviposition site, but also adjacent twig parts without eggs emit attractive volatiles (systemic effect). During egg deposition, *D. pini* females tangentially slit small egg pockets out of the pine needles with their sclerotized ovipositor valves, and insert the eggs into the wounds of the needles. Finally, eggs are covered on the top by a mixture of a frothy secretion and needle tissue that hardens within a few hours (Eliescu 1932). The elicitor inducing the pine's response was found to be located in the egg coating oviduct secretion of the female sawfly, since application of the oviduct secretion into artificially wounded pine needles also resulted in the induction of volatiles attractive to the egg parasitoid whereas artificial wounding alone did not (Hilker et al. 2002a). For application of the oviduct secretion into artificially wounded pine needles, oviducts were dissected from sawfly females and secretion was obtained by washing the oviducts in distilled water. The freshly isolated secretion was directly transferred into the wound of a pine needle (Hilker et al. 2002a).

Prior to the study presented here, nothing was known about the chemistry and stability of the elicitor located in the oviduct secretion. Thus, we tested whether the elicitor could be isolated at best by distilled water or other solvents and we examined how to store it. Furthermore, we did not know whether wounding of the pine needle is necessary for the eliciting process or whether the elicitor is also active when being applied onto an intact, non-wounded pine needle. Additionally, since our method to obtain oviduct secretion from sawfly females did not avoid contamination with hemolymph, the role of hemolymph for the eliciting activity of the oviduct secretion was unclear. Since oviducts and accessory reproductive structures are known to contain especially proteins (Gillot 2002; Hinton 1981), we examined whether a proteinase destroys the eliciting activity of the oviduct secretion.

**Material and methods**

*Plants and insects*

Branches of *Pinus sylvestris* used for experiments and rearing were detached from crowns of 15- to 35-year-old trees near Berlin. All stems were cleaned and sterilized according to the method of Moore and Clark (1968). *Diprion pini* was reared in the laboratory on pine twigs, as described by Bombosch and Ramakers (1976) and Eichhorn (1976) at 25±1 °C, L/D 18:6 h, and 70 % relative humidity. The egg parasitoid *Chrysonotomyia ruforum* was collected in the field in France (near Fontainebleau) and southern and central Finland. Parasitized eggs were kept in Petri dishes (i.d. 9 cm) in a climate chamber at 10 °C, 18L:6D photoperiod and 70 % relative humidity. To induce parasitoid emergence, needles with parasitized eggs were placed in a climate chamber at 25 °C, 18L:6D photoperiod and 70 % relative humidity. Emerging adults were collected daily and transferred in small perspex tubes (75mm long, 15mm i.d.) covered with gauze at one end. A cotton-wool ball moistened with an aqueous honey solution closed the other end. The parasitoids were kept at 10 °C, 18L:6D until they were used for bioassays. Female parasitoids used in the bioassays were experienced, i.e. they previously had contact with pine twigs carrying eggs of *D. pini* (Hilker et al. 2002a). Therefore, 48 h prior to the experiment, female and male parasitoids were brought into contact with a plant-host complex, consisting of a pine twig on which eggs of *D. pini* had been deposited, adult sawflies and a cotton-wool pad with aqueous honey solution. 24 h later, the parasitoids were removed from the plants and kept without any contact to the plant-host complex for another 24 h prior to the experiments.

*Olfactometer bioassay – general procedures and data collection*

All bioassays were conducted with a four-arm olfactometer as described in detail by Hilker et al. (2002b). The airflow was adjusted to 155 ml min$^{-1}$. We recorded how long the parasitoid was present within each of the four odour fields over a period of 600 s using the software program The Observer 3.0 (Noldus, Wageningen, The Netherlands). Only data obtained from active parasitoids walking at least 300 s of the observation period were used for statistical analyses (see below). Parasitoids preferably walking in the olfactometer field provided with the tested odour were defined to be "attracted" since significantly longer walking periods in the odour field is usually interpreted as a response of the parasitoid to an attractive odour (Hilker et al. 2002a). The number of parasitoids used for each bioassay was 22 to 36 (see Table 1). The odour source was changed after 5–9 parasitoids had been tested.

*Plant Treatments general*

Small pine twigs of about 20 cm length with ca. 90–120 needles were cut, placed into water, and treated for a period of 72 h at 25 °C and L/D 18 : 6. In all but one bioassay (see Table 1) artificially wounded pine twigs were used. For artificial wounding, pine needles were slit tangentially with a scalpel prior to treatment as described by Hilker et al. (2002b). If not mentioned otherwise, the differently treated oviduct secretion samples or respective control samples were applied into artificially wounded pine needles (see below). Eight needles of each twig were treated. All pine twigs were tested systemically as described in detail by Mumm et al. (2003), i.e. the lower half of a pine twig was treated, while the upper part was left untreated but was wrapped in a polyethylene terephthalate (PET) foil to avoid adsorption of volatiles from other parts of the twig. After 72 h, the upper half of the twig was cut off, the PET-foil was removed, the cut end of the twig was tightly wrapped with Parafilm®, and was transferred into the olfactometer to test for systemic induction. If the twig had been systemically induced, the parasitoids were expected to be attracted to its volatiles (Hilker et al. 2002a).

*I. Bioassays: How to solve and store the oviduct secretion ?*

(a) In a first experiment, the oviduct secretion was yielded by dissecting the oviducts (*oviductus lateralis and o. communis*) of 4 *D. pini* females. Oviducts were transferred in 8μl of ice-cold Aqua dest. To remove cell fragments of the oviducts, samples were centrifuged (10 min at 12,000 rpm) and the oviduct secretion containing supernatant was immediately applied in 1μl portions into wounded pine needles (i.e. 8 needles per twig). After 72 h, the untreated upper part of the pine twig was tested in the olfactometer.

(b) While in the first bioassay oviduct secretion was applied into wounded pine needles without longer storage between dissection and treatment, we examined whether oviduct secretion keeps its activity also after 3 h of storage at room temperature (ca. 20 °C). This storage period was chosen, since a proteinase K treatment (see below) needs an incubation time for 3 h at room temperature. Oviduct secretion was obtained, diluted in Aqua dest. and treated as described for bioassay (a). However, prior to application into the wounded pine needles, oviduct secretion was left in the dark at room temperature for a period of 3 h.

(c) To test whether oviduct secretion diluted in Aqua dest. keeps its activity when kept frozen directly after dissection, it was stored in a freezer for at least two days at −80 °C. Samples defrosted at room temperature within a few minutes and were then directly applied into wounded pine needles. Further conditions were the same as described for bioassay (a). Freezing the oviduct secretion without loosing its activity would be very useful to yield larger amounts that are necessary for further analyses.

(d) For control, slit pine needles were treated with Aqua dest. Eight needles were treated and 1 μl was applied per wounded needle.

(e) Oviduct secretion was obtained as described above (bioassay a), but transferred into Ringer solution (pH 7.2, Merck, Darmstadt, Germany) and then stored in the dark for 3 h at room temperature (ca. 20°C). Treatment of pine needles was the same as described for bioassay (a).

(f) To test whether oviduct secretion diluted in Ringer solution can be stored frozen without loosing activity, we kept it frozen as described for bioassay (c). For the bioassay, defrosted oviduct secretion in Ringer solution was directly applied into wounded pine needles.

(g) For control, eight wounded pine needles were treated with Ringer solution (1 µl per needle) only.

*II. Bioassay: Is damage of needles necessary for the eliciting effect of oviduct secretion?*

(h) Oviduct secretion was obtained as described for bioassay (a) and transferred into Ringer solution (pH 7.2, Merck, Darmstadt, Germany). However, pine needles were not wounded prior to application of the secretion. Instead, 1 µl secretion was slowly applied on an intact needle.

*III. Bioassay: Does hemolymph have eliciting activity?*

(i) Legs of sawfly females were cut and 1 µl of the emerging hemolymph was yielded with a glass capillary and transferred into Ringer solution. After 3 h storage in the dark at room temperature, slit pine needles were treated with these samples (compare bioassay (e)).

*IV. Bioassay: Does a proteinase K destroy the activity of the elicitor?*

(j) Oviduct secretion was obtained and stored in Ringer solution as described for bioassay (e). 10 vol. % Proteinase K (pH 7.2; Merck, Darmstadt, Germany) was added to the sample. During an incubation period of 3 h, samples were stored in the dark at room temperature (ca. 20 °C). The proteinase K was covalently bound to small latex beads. Therefore, it could be separated from the secretion after incubation by centrifuging the sample for 10 min at 10,000 rpm. The supernatant was then applied in artificially wounded needles as described above.

*Sodium Dodecyl Sulfate–Polyacrylamide Gel Electrophoresis (SDS–PAGE)*

In order to examine the digestion of proteins of the oviduct secretion by proteinase K treatment and to check that all proteinase K had been removed prior to the treatment of needles, SDS-PAGE of digested and undigested oviduct secretion was performed. Samples were prepared as described for bioassay (e) and (j). Digested samples were analysed after removal of proteinase K by centrifugation. For comparison, also hemolymph samples were analysed by SDS-PAGE. Electrophoresis was performed on a gradient gel (T= 5-17.5%; C=4%) according to the procedure of Laemmli (1970). Protein molecular mass markers (Precision Plus Protein, BioRad, München, Germany) were used. Electrophoresis was started with 100 V until a uniform front occurred. Then, the voltage was raised to 200 V. The gels were stained with Coomassie Brilliant Blue R-250 (Roth, Germany). To obtain a better visualization of the bands, the gels were analysed by using the software program Scion Image (Scion Corp., Frederick, MY, USA).

*Statistics*

Bioassay data were statistically analysed by using the Friedman analysis of variance (ANOVA) for comparing residence time within each of the four olfactometer fields using the software program SPSS 11.0. (SPSS Inc., USA). The Wilcoxon-Wilcox test was used for post-hoc comparisons (Köhler et al. 1995).

**Results**

*Bioassays*

The egg parasitoid *C. ruforum* was significantly attracted to volatiles from a pine twig that was treated with oviduct secretion of *D. pini* when the freshly obtained oviduct secretion had been directly applied to the needles without storage (Table 1, bioassay a). Volatiles attractive to the parasitoid were released by the upper part of the twigs adjacent to the lower parts that had been treated. Thus, application of oviduct secretion into wounded pine needles is not only able to induce a local response, but also a systemic one confirming the results by Hilker et al. (2002a). The activity of the elicitor within the oviduct secretion was not stable and got lost, when the secretion was diluted in Aqua dest. and then stored for 3 h or kept frozen at – 80 °C (Table 1, bioassay b and c). However, when oviduct secretion was diluted in Ringer solution, its activity was stable both after storage at room temperature for 3 h and after freezing at –80 °C (Table 1, bioassay e and f). The control experiments revealed that neither Aqua dest. nor Ringer solution itself have any eliciting effect (Table 1, bioassay d and g).

Wounding of pine needles revealed to be essential for the induction process. Volatiles from undamaged pine twigs treated only topically with oviduct secretion did not attract *C. ruforum* (Table 1, bioassay h). Therefore, the elicitor is only active when transferred into damaged needle tissue.

Hemolymph of *D. pini* females did not induce the emission of pine volatiles attractive for the egg parasitoids (Table 1, bioassay i). Since the oviduct secretion naturally does not come into contact with hemolymph during oviposition, free hemolymph in the female's abdomen does *per se* obviously not play a role for the volatile induction by egg deposition.

The activity of the elicitor in the oviduct secretion was lost after treatment with proteinase K, because volatiles of pine twigs treated with oviduct secretion after proteinase K digestion were not attractive to the parasitoids (Table 1, bioassay j).

**Table 1** Response of female egg parasitoids *C. ruforum* to volatiles released systemically from differently treated *Pinus sylvestris* twigs. The time parasitoid females spent in test and control fields (contr. 1 –3) of a four-arm-olfactometer are given. Test field with odour from differently treated pine; Contr. 1, 2, 3, = three fields with control (clean) air. Column "oviduct secretion" indicates whether the pine twig has been treated with oviduct secretion (+) or not (none). The columns "solvent" and "treatment" indicate how the oviduct secretion and hemolymph, respectively, have been treated prior to application to the pine twig. Median and interquartile range (parentheses) are given. * indicates a significant (p<0.05), and n.s. a non-significant (p>0.05) difference evaluated by a Friedman ANOVA. Different letters indicate significant (p<0.05) differences evaluated by the Wilcoxon–Wilcox test.

| Bio-assay | Oviduct secretion | Solvent | Treatment | Application into | residence time [s] | | | | N | Statistics |
|---|---|---|---|---|---|---|---|---|---|---|
| | | | | | Test | Contr. 1 | Contr. 2 | Contr. 3 | | |
| **I. How to solve and store the elicitor ?** | | | | | | | | | | |
| a | + | Aqua dest. | freshly obtained | slit pine needles | 221[a] (43-447) | 97[ab] (4-181) | 53[b] (1-135) | 49[ab] (5-208) | 36 | * (p=0.02) |
| b | + | Aqua dest. | kept at room temperature (3h) | slit pine needles | 181 (56-292) | 125 (52-250) | 83 (18-209) | 90 (46-135) | 32 | n.s. (p>0.05) |
| c | + | Aqua dest. | after freezing at –80°C | slit pine needles | 120 (64-248) | 146 (48-208) | 124 (61-197) | 143 (46-204) | 31 | n.s. (p>0.05) |
| d | none | Aqua dest. | -- | slit pine needles | 83 (19-168) | 69 (27-188) | 194 (49-410) | 94 (8-239) | 22 | n.s. (p>0.05) |
| e | + | Ringer solution | kept at room temperature (3h) | slit pine needles | 184[a] (91-351) | 99[ab] (58-187) | 58[b] (4-160) | 115[ab] (24-210) | 26 | * (p=0.02) |
| f | + | Ringer solution | after freezing at –80°C | slit pine needles | 198[a] (131-307) | 95[ab] (21-157) | 55[b] (17-160) | 128[ab] (55-248) | 25 | * (p=0.02) |
| g | none | Ringer solution | -- | slit pine needles | 92 (28-201) | 96 (14-200) | 145 (63-309) | 125 (55-183) | 27 | n.s. (p>0.05) |
| **II. Is damage of needles necessary for the eliciting effect ?** | | | | | | | | | | |
| h | + | Ringer solution | freshly obtained | undamaged needles | 74 (39-150) | 141 (67-232) | 166 (95-249) | 127 (67-214) | 28 | n.s. (p>0.05) |
| **III. Does hemolymph have eliciting activity ?** | | | | | | | | | | |
| i | none | Ringer solution | hemolymph | slit pine needles | 158 (48-283) | 123 (52-193) | 97 (29-227) | 104 (63-202) | 31 | n.s. (p>0.05) |
| **IV. Does a proteinase K destroy the activity of the elicitor ?** | | | | | | | | | | |
| j | + | Ringer solution | proteinase K treatment (3h) | slit pine needles | 74 (23-143) | 62 (17-223) | 128 (59-248) | 107 (38-315) | 27 | n.s. (p>0.05) |

SDS-PAGE

Electrophoresis of undigested oviduct secretion showed 7 bands with molecular masses from ca. 10-250 kDa, visualized by 7 peaks in the Scion image. The band pattern of the hemolymph was similar, but one band (No. 1) was missing compared to the oviduct secretion. SDS-gels of digested oviduct secretion showed no bands, confirming that all proteins were destroyed by proteinase K treatment. Furthermore, no proteinase K residues were detected here, thus confirming that the enzyme was completely removed from the sample prior to its application to pine needles (Fig. 1).

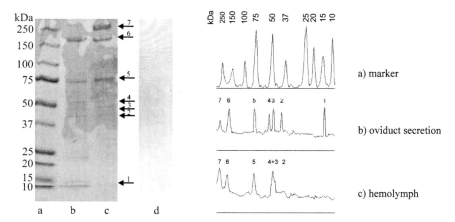

**Figure 1** <u>Left</u>: SDS-PAGE after staining with Coomassie brilliant blue. (a) marker, (b) oviduct secretion undigested, (c) hemolymph, and (d) oviduct secretion digested by proteinase K. <u>Right</u>: Scion image plot of the same SDS-gel showing gel bands as peaks. Numbers 1-7 indicate bands detected on the gel.

**Discussion**

Our results clearly show that the oviduct secretion of *D. pini* contains an elicitor which induces a systemic release of pine volatiles attractive to the egg parasitoid *C. ruforum*. This conforms the results by Hilker et al. (2002b), giving further support that oviposition and oviduct secretion induce similar response in *P. sylvestris*. The elicitor is unstable in Aqua dest., but can be stabilized by Ringer solution and its digestion by proteinase K indicates that the elicitor is a peptide or protein or a compound bound to a peptide or protein (Falbe and Regitz 1990).

To our knowledge, only two elicitors inducing plant responses to egg depositions have been intensively investigated so far with respect to their chemistry (Hilker et al. 2002b):

(a) In bruchid beetles, so-called bruchins have been isolated as components eliciting a response in peas (*Pisum sativum* L.) that directly affects the herbivore. Egg depositions of the beetles induce growth of neoplasms at the site of egg attachment. The plant's response to egg deposition may protect the pea pod from larval feeding damage, because the egg on the neoplasm may easily be detached from the pod (Doss et al. 1995, 2000). When isolated bruchins are applied onto responsive pea pods, they also induce neoplastic growth as observed after egg deposition. The bruchins are long-chain $\alpha,\omega$-monounsaturated $C_{22}$-diols and $\alpha,\omega$-mono- and diunsaturated $C_{24}$-diols, (Oliver et al. 2000, 2002).

(b) In the tenthredinid sawfly *Pontania proxima*, an elicitor of a plant reaction is known to be located in the secretion of accessory glands. When *P. proxima* lays an egg onto a *Salix fragilis* leaf, the egg deposition induces mitogenesis of plant tissue and growth of a gall is initiated. Chemical analyses revealed that the secretion of the accessory glands contains nucleic acids, protein (Hovanitz 1959), uric acid, adenine derivatives, glutamic acid, and possibly uridine (McCalla et al. 1962). Leitch (1994) suggested that precursors of cytokinins may be present in the ovipositional fluid and act as gall-initiators, however, Higton and Mabberley (1994) doubt that cytokinins *induce* the galls.

Several elicitors of plant defensive responses to feeding herbivores have been isolated from regurgitate of herbivorous larvae (Felton and Eichenseer 1999). The components are known to be proteinous or to contain a peptide bond. $\beta$-Glucosidase has been isolated from the regurgitate of *Pieris brassicae* larvae (Mattiacci et al. 1995). Volicitin (*N*-[17-hydroxylinolenoyl]-L-glutamine) and other fatty acid – amino acid conjugates have been isolated from the regurgitate of several lepidopteran species (Alborn et al. 1997, 2000; Halitschke et al. 2001; Mori et al. 2001; Pohnert et al. 1999; Turlings et al. 2000). Musser et al. (2002) isolated a glucose oxidase from the saliva of *Heliothis zea*. Their results indicate that release of this glucose oxidase elicits a suppression of the plant's defensive response, and thus, has been interpreted as counteradaptation by the herbivore. Little is known about the mode of action of herbivore elicitors (Dicke and van Poecke 2002). In contrast to elicitors isolated from plant pathogens, no herbivore defence-related plant perception mechanisms have been identified so far (Dicke and van Poecke 2002; Felton and Eichenseer 1999; Ham and Bent 2002; Martin et al. 2003).

Our results revealed that the elicitor present in the oviduct secretion of *D. pini* needs to come into contact with wounded plant tissue. Similar results were found by Meiners and Hilker (2000) in another tritrophic system consisting of the elm *Ulmus minor*, the leaf beetle *Xanthogaleruca luteola*, and the egg parasitoid *Oomyzus gallerucae*. The elicitor present in the oviduct secretion of the elm leaf beetle also needs to contact wounded elm tissue to induce production of volatiles which attract the egg parasitoid of the elm leaf beetle. Elicitors of feeding herbivores inducing plant responses have also been shown to need disrupted plant tissue to become active (Mattiacci et al. 1995, Turlings et al. 1990). In contrast, Colazza et al. (2004a) could show that the pentatomid *Nezara viridula* lays its eggs on bean leaves without wounding the plant tissue. These egg depositions induced the bean leaves to release volatiles that attract the egg parasitoid. However, bean leaves carrying eggs of the bug only released volatiles attractive for the parasitoid, when leaves had also been damaged by feeding. Leaves with eggs, but without feeding damage did not emit the attractive volatiles. Thus, also in this tritrophic system, leaf damage seems to be necessary to induce the volatile blend attractive for the parasitoid. However, the elicitor associated with the egg deposition seems to be able to become active also when being applied to intact leaf surface.

## Acknowledgements

We are grateful to Lars Podsiadlowski (FU Berlin, Institut für Biologie, Systematik und Evolution der Tiere) for assistance with SDS-PAGE. Special thanks also to Wolfgang Groß (FU Berlin, Institut für Biologie, Pflanzenphysiologie) for his assistance and many helpful comments during this project. This project was funded by the Deutsche Forschungsgemeinschaft (Hi 416 11-1/2).

## References

Alborn, HT, Jones, TH, Stenhagen, G, Tumlinson, JH (2000). Identification and synthesis of volicitin and related components from beet armyworm oral secretions. J. Chem. Ecol. 26:203-220.

Alborn, HT, Turlings, TCJ, Jones, TH, Stenhagen, G, Loughrin, JH, Tumlinson, JH (1997). An elicitor of plant volatiles from beet armyworm oral secretion. Science 276:945-949.

Baldwin, IT, Preston, CA (1999). The eco-physiological complexity of plant responses to insect herbivores. Planta 208:137-145.

Boland, W, Koch, T, Krumm, T, Piel, J, Jux, A (1999). Induced biosynthesis of insect semiochemicals in plants. In: Chadwick, DJ, Goode, JA (eds), Insect-Plant Interactions and Induced Plant Defence. John Wiley & Sons Ltd, Chicester, pp. 110-126.

Bombosch, S, Ramakers, PMJ (1976). Zur Dauerzucht von *Gilpinia hercyniae* Htg. Z. Pflanzenkrank. Pflanzen. 83:40-44.

Colazza, S, Fucarino, A, Peri, E, Salerno, G, Conti, E, Bin, F (2004a). Insect oviposition induces volatile emission in herbaceous plants that attracts egg parasitoids. J. Exp. Biol. 207:47-53.

Colazza, S, McElfresh, JS, Millar, JG (2004b). Identification of volatile synomones, induced by *Nezara viridula* feeding and oviposition on bean spp., that attract the egg parasitoid *Trissolcus basalis*. J. Chem. Ecol. 30 :945-964.

Dicke, M, Hilker, M (2003). Induced plant defences: from molecular biology to evolutionary ecology. Basic Appl. Ecol. 4 :3-14.

Dicke, M, van Poecke, RMP (2002). Signalling in plant-insect interactions: signal transduction in direct and indirect plant defence. In: Scheel, D , Wasternack, C (eds), Plant Signal Transduction. Oxford University Press, Oxford, pp. 289-316.

Dicke, M, van Loon, JJA (2000). Multitrophic effects of herbivore-induced plant volatiles in an evolutionary context. Entomol. Exp. Appl. 37:237-249.

Doss, RP, Oliver, JE, Proebsting, WM, Potter, SW, Kuy, SR, Clement, SL, Williamson, RT, Carney, JR, DeVilbiss, ED (2000). Bruchins: Insect derived plant regulators that stimulate neoplasm formation. Proc. Natl. Acad. Sci. USA 97:6218-6223.

Doss, RP, Proebsting, WM, Potter, SW, Clement, SL (1995). Response of *Np* mutant of pea (*Pisum sativum* L.) to pea weevil (*Bruchus pisorum* L.) oviposition and extracts. J. Chem. Ecol. 21:97-106.

Eichhorn, O (1976). Dauerzucht von *Diprion pini* L. (Hym.: Diprionidae) im Laboratorium unter Berücksichtigung der Fotoperiode. Anz. Schädlingskd. Pfl. 49:38-41.

Eliescu, G (1932). Beiträge zur Kenntnis der Morphologie, Anatomie und Biologie von *Lophyrus pini* . Z. ang. Ent. 19:22 (188)-67 (206).

Falbe, J, Regitz, M (1990). Römpp - Chemie Lexikon. Thieme, Stuttgart.

Felton, GW, Eichenseer, H (1999). Herbivore saliva and its effects on plant defense against herbivores and pathogens. In: Agrawal, AA, Tuzun, S, Bent, E (eds), Induced Plant Defenses against Pathogens and Herbivores. APS Press, St. Paul, pp. 19-36.

Gillott, C (2002). Insect accessory reproductive glands: Key players in production and protection of eggs. In: Hilker, M, Meiners, T (eds), Chemoecology of Insect Eggs and Egg Deposition. Blackwell Publishing, Berlin, Oxford, pp. 37-59.

Halitschke, R, Schittko, U, Pohnert, G, Boland, W, Baldwin, IT (2001). Molecular interactions between the specialist herbivore *Manduca sexta* (Lepidoptera, Sphingidae) and its natural host N*icotiana attenuat*a. III. Fatty acid-amino acid conjugates in herbivore oral secretions are necessary and sufficient for herbivore-specific plant responses. Plant Physiol. 125:711-717.

Ham, JH, Bent, A (2002). Recognition and defence signalling in plant/bacterial and fungal interactions. In: Scheel, D, Wasternack, C (eds), Plant Signal Transduction. Oxford University Press, Oxford, pp. 198-225.

Higton, RN, Mabberley, DJ (1994). A willow gall from the galler's point of view. In: Williams, MAJ (ed), Plant Galls, Organisms, Interactions, Populations. Oxford University Press, Oxford, pp. 301-312.

Hilker, M, Meiners, T (2002). Induction of plant responses towards oviposition and feeding of herbivorous arthropods: a comparison. Entomol. Exp. Appl. 104:181-192.

Hilker, M, Kobs, C, Varama, M, Schrank, K (2002a). Insect egg deposition induces *Pinus* to attract egg parasitoids. J. Exp. Biol. 205:455-461.

Hilker, M, Rohfritsch, O, Meiners, T (2002b). The plant's response towards insect egg deposition. In: Hilker, M, Meiners, T (eds), Chemoecology of Insect Eggs and Egg Deposition. Blackwell Publishing, Berlin, Oxford, pp. 205-233.

Hinton, HE (1981). Biology of Insect Eggs. Pergamon Press, Oxford.

Hovanitz W. (1959). Insects and plant galls. Sci. Am. 201:151-162.

Köhler, W, Schachtel, G, Voleske, P (1995). Biostatistik. Springer, Berlin.

Laemmli, UK (1970). Cleavage of structural proteins during the assembly of the head of bacteriophage T4. Nature 227:680-685.

Leitch, IJ (1994). Induction and development of bean galls caused by *Potonia proxima*. In: Williams, MAJ (ed), Plant Galls, Organisms, Interactions, Populations. Oxford University Press, Oxford, pp. 283-300.

Martin, GB, Bogdanove, AJ, Sessa, G (2003). Understanding the functions of plant disease resistance proteins. Annu. Rev. Plant Biol. 54:23-61.

Mattiacci, L, Dicke, M, Posthumus, MA (1995). β-Glucosidase: An elicitor of herbivore-induced plant odor that attracts host-searching parasitic wasps. Proc. Natl. Acad. Sci. USA 92:2036-2040.

McCalla, DR, Genthe, MK, Hovanitz, W (1962). Chemical nature of insect gall growth factor. Plant Physiol. 37:98-103.

Meiners, T, Hilker, M (2000). Induction of plant synomones by oviposition of a phytophagous insect. J. Chem. Ecol. 26:221-232.

Moore, GE, Clark, EW (1968). Suppressing microorganisms and maintaining turgidity in coniferous foliage used to rear insects in the laboratory. J. Econ. Entomol. 61:1030-1031.

Mori, N, Alborn, HT, Teal, PEA, Tumlinson, JH (2001). Enzymatic composition of elicitors of plant volatiles in *Heliothis virescens* and *Helicoverpa zea*. J. Insect Physiol. 47:749-757.

Mumm, R, Schrank, K, Wegener, R, Schulz, S, Hilker, M (2003). Chemical analysis of volatiles emitted by *Pinus sylvestris* after induction by insect oviposition. J. Chem. Ecol. 29:1235-1252.

Musser, RO, Hum-Musser, SM, Eichenseer, H, Pfeiffer, M, Ervin, G, Murphy, JB, Felton, GW (2002). Caterpillar saliva beats plant defences. Nature 116:599-600.

Oliver, JE, Doss, RP, Marquez, B, DeVilbiss, ED (2002). Bruchins, plant mitogens from weevils: structural requirements for activity. J. Chem. Ecol. 28:2503-2513.

Oliver, JE, Doss, RP, Williamson, RT, Carney, JR, DeVilbiss, ED (2000). Bruchins - mitogenic 3-(hydroxypropanoyl) esters of long chain diols from weevils of the Bruchidae. Tetrahedron 56:7633-7641.

Pohnert, G, Jung, V, Haukioja, E, Lempa, K, Boland, W (1999). New fatty acid amides from regurgitant of lepidopteran (Noctuidae, Geometridae) caterpillars. Tetrahedron 55:11275-11280.

Turlings, TCJ, Gouinguené, S, Degen, T, Fritzsche-Hoballah, ME (2002). The chemical ecology of plant-caterpillar-parasitoid interactions. In: Tscharntke, T, Hawkins, BA (eds), Multitrophic Level Interactions . Cambridge University Press, Cambridge, pp. 148-173.

Turlings, TCJ, Alborn, HT, Loughrin, JH, Tumlinson, JH (2000). Volicitin, an elicitor of maize volatiles in oral secretion of *Spodoptera exigua*: Isolation and bioactivity. J. Chem. Ecol. 26:189-202.

Turlings, TCJ, Tumlinson, JH, Lewis, WJ (1990). Exploitation of herbivore-induced plant odors by host-seeking parasitic wasps. Science 250:1251-1253.

# Chapter 3

# Chemical Analysis of Volatiles Emitted by *Pinus sylvestris* after Induction by Insect Oviposition

**Abstract.** GC-MS analyses of the headspace volatiles of Scots pine (*Pinus sylvestris*) induced by egg deposition of the sawfly *Diprion pini* were conducted. The odour blend of systemically oviposition-induced pine twigs, attractive for the eulophid egg parasitoid *Chrysonotomyia ruforum*, was compared to volatiles released by damaged pine twigs (control) which are not attractive for the parasitoid. The mechanical damage inflicted to the control twigs mimicked the damage conducted by a sawfly female prior to egg deposition. The odour blend released by oviposition-induced pine twigs consisted of numerous mono- and sesquiterpenes, which all were also present in the headspace of the artificially damaged control twigs. A quantitative comparison of the volatiles from oviposition-induced twigs and the controls revealed that only the amounts of (*E*)-β-farnesene were significantly higher in the volatile blend of the oviposition-induced twigs. Volatiles from pine twigs treated with jasmonic acid (JA) also attract the egg parasitoid. No qualitative differences were detected when comparing the composition of the headspace of JA-treated pine twigs with the volatile blend of untreated control twigs. JA-treated pine twigs also released significantly higher amounts of (*E*)-β-farnesene. However, the JA-treatment additionally induced a significant increase of the amount of further terpenoid components. The release of terpenoids by pine after wounding, egg deposition and JA-treatment is discussed with special respect to (*E*)-β-farnesene.

**Keywords.** egg deposition, Diprionidae, Eulophidae, induction, jasmonic acid, monoterpenes, parasitoids, *Pinus sylvestris,* sawfly, sesquiterpenes, systemic effect, terpenoids.

**Introduction**

Plants are able to respond to herbivore damage with direct or indirect defences (overviews in Karban and Baldwin 1997; Agrawal et al. 1999; Dicke 1999; Dicke and van Loon 2000; Walling 2000; Kessler and Baldwin 2002). While direct defences act directly against the herbivore by e.g. production of toxins or repellents (Baldwin 1994; Duffey and Stout 1996; Edwards and Wratten 1987), indirect plant defences employ the third trophic level, the carnivores (Price et al. 1980, Price 1986). Numerous studies show that a plant can attract predators and parasitoids of the herbivores by release of volatiles in response to herbivore attack (e.g. Dicke et al. 1990; Turlings et al. 1990; De Moraes et al. 1998; Dicke and Vet 1999). This induction of volatiles results in a change of plant odour that may be a quantitative alteration (i.e., the volatiles released by the undamaged and damaged plants are the same but are released in differing amounts) or a qualitative one (i.e., the damaged plant produces components that are not emitted by the undamaged plant) (e.g. Boland et al. 1999; Dicke 1999; Paré et al. 1999). The herbivore-induced change of a plant´s odour can be specific both for the plant and herbivore species, for the plant´s age as well as for the attacking herbivore stage (e.g. Takabayashi et al. 1994; De Moraes et al. 1998; Stout and Bostock 1999). The release of induced volatiles is often not restricted to the area of feeding damage, but the signal may be transported systemically to undamaged plant parts (e.g. Turlings et al. 1993; Dicke 1994; Karban and Baldwin 1997).

Several wound signals are involved in induced plant defences (Aducci 1997; Staswick and Lehman 1999, and references therein). Jasmonates have intensively been studied in context of plant defence against herbivores (Sembdner and Parthier 1993; Creelman and Mullet 1997; Beale and Ward 1998). Exogenous application of jasmonic acid (JA) or methyl jasmonate to undamaged plants was found to induce a quantitative and qualitative change of the plant´s odour that is similar to the change of volatiles induced by infestation with herbivores (Hopke et al. 1994; Boland et al. 1995, 1999; Koch et al. 1999; Ozawa et al. 2000). Even though JA- induced plant volatiles were found to be attractive to predators and parasitoids (Dicke et al. 1999;

Gols et al. 1999; Thaler 1999), some carnivores are able to distinguish between odour from herbivore and JA- induced plants (Dicke et al. 1999).

Apart from feeding, egg deposition by herbivores can also induce plants to emit volatiles that attract specialized egg parasitoids. This was shown in two different tritrophic systems:

(1) Meiners and Hilker (1997) demonstrated that the eulophid egg parasitoid *Oomyzus gallerucae* is attracted by volatiles of the field elm (*Ulmus minor*) induced locally and systemically by oviposition of the elm leaf beetle *Xanthogaleruca luteola*. The emission of volatiles attractive to the egg parasitoid is also inducible by application of JA (Meiners and Hilker 2000). Analyses of volatiles emitted by elm leaves revealed that egg deposition induces the emission of a complex odour blend qualitatively and quantitatively different from the odour of undamaged elm leaves. Treatment of elm leaves with JA also results in a qualitative and quantitative change of the plant´s odour which almost exclusively consists of terpenoids (Wegener et al. 2001).

(2) In the second tritrophic system, oviposition by the herbivorous sawfly *Diprion pini* (Hymenoptera, Diprionidae) induces Scots pine (*Pinus sylvestris*) to produce volatiles that attract the egg parasitoid *Chrysonotomyia ruforum* (Hymenoptera, Eulophidae). Prior to oviposition, a female of *D. pini* slits the pine needle with its ovipositor. Mimicking this mechanical damage with a razor blade does not induce pine twigs to emit attractive volatiles. The elicitor is located in the oviduct secretion that envelopes the sawfly eggs, since application of oviduct secretion (without eggs) elicits the release of attractive volatiles. The oviposition-induced plant response is systemic, i.e. volatiles emitted from needles adjacent to the oviposition sites are attractive towards the egg parasitoids 72 hours after egg deposition. As has been shown in the elm system, exogenous application of JA results in an induction of attractive volatiles (Hilker et al. 2002a,b; Hilker and Meiners 2002).

In the present study, we conducted GC-MS analyses of the headspace of *P. sylvestris* induced systemically by egg deposition of *D. pini*. In order to detect the volatiles that attract the egg parasitoid, the attractive odour blend of systemically oviposition-induced pine was compared to the non-attractive odour released by artificially damaged pine twigs mimicking the slitting conducted by a *D. pini* female prior to oviposition. In addition, the odour blend of attractive JA-treated pine twigs was analysed.

## Materials and methods

*Plants and insects*

All branches of P. sylvestris used for analyses were detached from crowns of 35 - 45 year old trees in the forests near Berlin. The lower part of the stems was cleaned and sterilized according to Moore and Clark (1968). D. pini was reared in the laboratory as described by Bombosch and Ramakers (1976) and Eichhorn (1976) at 25 °C, L18 : D6 and 70 % relative humidity.

*Plant treatments general*

Small pine twigs with about 90 to 120 needles were cut, placed into water and treated for a period of 72 hrs at 25 °C, L18 : D6 and approx. 2000 lux. All treatments were conducted as described in detail by Hilker et al. (2002a). Headspace samples were taken from the twigs immediately after treatment (see below).

*Systemically Oviposition-Induced Pine Twig*

Females of *D. pini* were offered the lower half of a pine twig for egg deposition, while the upper part was wrapped in a polyethylene terephthalate (PET) foil to avoid egg deposition and adsorption of volatiles from other parts of the twig or sawflies. This ensured that collected volatiles originated from plants only. The bag was ventilated with purified air through an in- and outlet. When *D. pini* females had laid at least eight egg clusters after 72 h on the lower part of the pine twig, the foil wrapping the upper half without eggs was removed. The upper half of the twig was cut off and volatiles from this systemically oviposition-induced half of the pine twig were collected.

*Artificially damaged Pine twig*

This treatment served as control for the systemically oviposition-induced twigs. Because terpenoid concentrations in pine species are well-known to show a high genotypic variability (e.g. Petrakis et al. 2001) and even vary within an individual tree (e.g. Gershenzon and Croteau 1991; Manninen et al. 2002), the control twigs were taken from the same large pine branch as the twigs for the oviposition-induction treatment and were cut right next to them. The control twigs were kept under the same conditions as the oviposition-induced ones, but 10 needles of the lower half of a pine twig were slit with a scalpel to mimic the mechanical damage that a *D. pini* female conducts with her ovipositor prior to oviposition. Headspace samples were taken from the upper, undamaged half of this pine twig mechanically damaged in its lower half.

*Treatment of Pine twigs with Jasmonic Acid*

Pine twigs were supplied with 0.3 µmol / ml jasmonic acid (JA) dissolved in an aqueous Tween 20 solution (0.05 %). This concentration of JA induces the emission of attractive volatiles for the egg parasitoid *C. ruforum* (Hilker et al. 2002a). Control twigs were provided with Tween 20 solution only and again taken from the same branch right next to the ones used for the JA-treatment.

*Collection of headspace volatiles and internal standard*

Volatile compounds emitted by the pine twigs were collected using a dynamic headspace sampling method. A control twig that was compared with an oviposition-induced pine twig was cut from the same large branch as the twig for induction by oviposition was taken from. Also a control that was compared with a JA-treated pine twig was taken from the same large branch as the twig for JA-treatment. Thus, test and control twigs may be considered as paired samples (see below: statistical analyses). The cut end of a twig was carefully and tightly wrapped with Parafilm® during the sampling period of five hours. To collect volatiles from test and control twigs, the air was initially purified by passing over activated charcoal, then it was split before entering the flasks (250 ml) containing either the test or the control samples. The volatiles of each sample were collected separately on 5 mg charcoal filters and were eluted with 50 µl dichloromethane containing 125 ng / µl methyl octanoate as internal standard. Two flowmeters (Supelco, Germany) ensured a constant airflow of 110 ml/ min. All components of the sampling set-up were connected with teflon tubing. The conditions during sampling were the same as those used for the treatments (see above).

*Chemical analysis of headspace samples*

Headspace samples were analysed by coupled gas chromatography – mass spectrometry (GC-MS) on a Fisons GC model 8060 and a Fisons MD 800 quadrupole MS using a J&W 30 m DB5-ms capillary column (0.32 mm internal diameter, film thickness 0.25 µm). The samples (1 µl) were injected in splitless mode (injector temperature 240 °C) with helium as the carrier gas (inlet pressure 10 kPa). The temperature program started at 40 °C (4 min hold) and rose with 3 °C/min up to 280 °C. The column effluent was ionized by electron impact ionization (EI) at 70 eV. Mass range was from 35 - 350 *m/z* with a scan time of 0.9 s and an interscan delay of 0.1 s. The eluted compounds were identified by comparing the mass spectra and linear retention indices with those of authentic reference compounds or with NIST library spectra. Retention indices were calculated for each compound according to van den Dool and Kratz (1963) and compared to critically evaluated tabulated data (Adams 1995; Joulain and König 1998).

*Enantiomer separation*

Enantiomer separations of several chiral monoterpenes were performed on a Hewlett Packard GC 6890 coupled to a MSD 5973, equipped with a 35 m capillary column (id = 0.25 mm) coated with hydrodex-6-TBDMS (Macherey-Nagel, Düren, Germany). Helium was used as the carrier gas at a flow rate of 1 ml/min. The temperature of the gas chromatograph was set at 50 °C. The injection port temperature was 200 °C. The samples were injected in splitless mode. The mass spectrometer was operated in EI ionization mode (70 eV).

The enantiomers of limonene, α-pinene, and β-pinene were obtained from Fluka (Taufkirchen, Germany). (+)-Sabinene, which contained minor amounts of (–)-

sabinene, was purchased from Carl Roth GmbH & Co KG (Karlsruhe, Germany), while racemic and (−)-camphene were obtained from Sigma-Aldrich (Taufkirchen, Germany) as was (+)-3-carene. The enantiomer (−)-3-carene, which is a major component of the essential oil of black pepper (König et al. 1992), was obtained by us by steam distillation of a commercially available sample (coarse variety, ALDI, Germany). While (−)-β-phellandrene occurs in *Pinus sylvestris* (Borg-Karlson et al. 1993; Sjödin et al. 1996), (+)-β-phellandrene was obtained by SPME sampling of leaves of lovage (*Levisticum officinale* Koch.), in which it was the major component (König et al. 1992). 2-Carene is known to occur naturally only as the (+)-enantiomer (Fugmann et al. 1997). All the mentioned enantiomers could be separated on the hydrodex column (see above). The only problem occurred with (+)-limonene, which eluted in the beginning of the (−)-β-phellandrene peak. However, careful analysis of the ion traces of the characteristic ions of these compounds [(+)-limonene $m/z = 68$, (−)-β-phellandrene $m/z = 93$] allowed determination of their relative amounts. The retention times (in minutes) of the compounds were as follows: (−)-α-pinene (21.0), (+)-α-pinene (22.3), (−)-camphene (24.6), (+)-camphene (26.8), (+)-sabinene (27.3), (+)-β-pinene (29.9), (−)-sabinene (30.3), (−)-β-pinene (31.8), (+)-3-carene (34.0), (−)-3-carene (35.3), (−)-limonene (40.9), (+)-limonene (44.4), (−)-β-phellandrene (44.5), (+)-β-phellandrene (45.7).

*Statistical analysis*

Headspace samples were taken from 10 oviposition-induced twigs and 10 respective controls (artificial damage), as well as from 10 JA-treated twigs and 10 respective controls. A calibration with a set of pure standards and the internal standard has been performed to calculate the response factors. Peak areas of the components compared to the peak areas of the internal standard were quantified with respect to the response factors. Limonene and β-phellandrene were not separated on the DB-5 column. Analogous to the enantiomeric separation of these two components careful analysis of the area of the ion traces ($m/z = 68$ for limonene and $m/z = 93$ for β-phellandrene) with respect to the portion of these ions of the total ion current allowed the determination of their relative quantities (compare formula given in Table 1 for details of calculation). The relative peak areas of test and control twigs were statistically compared as paired samples by the Wilcoxon matched pairs test using the software program SPSS (SPSS Inc.). Ratios of enantiomers in test and control twigs were also compared by the Wilcoxon matched pairs test. Numbers of samples analysed quantitatively differ from the number of samples on which enantiomer separation was performed, because the quantitative analyses were conducted first, followed by enantiomer separation of the remaining samples by use of a different equipment.

**Results**

The compounds detected in at least five out of 10 samples of the oviposition-induced pine twigs are listed in Table 1. A characteristic total ion current chromatogram of the headspace of an oviposition induced pine twig is shown in Figure 1. All components detected in the headspace of oviposition-induced pine twigs and the JA-treated ones were also found in the respective controls, i.e. no qualitative change in the headspace of pine twigs occurred due to oviposition or JA-treatment. The pine twigs emitted predominantly mono- and sesquiterpenes regardless of their treatment. The major compounds in the headspace of oviposition-induced pine twigs were myrcene, 3-carene, and β-phellandrene (at least 10% of total peak area, Table 1, Figure 1). These three compounds were also dominant in the headspace of JA-treated pine twigs, but here additionally α-pinene and β-pinene occurred as major components.

The volatile pattern of JA-treated pine twigs was not as consistent as in oviposition-induced twigs since only 9 out of 31 compounds could be detected in the headspace of JA-treated pine twigs as often as in oviposition-induced pine twigs (Figure 2; see numbers written in columns for components no. 2, 3, 6, 21, 24, 26, 27, 30, 31). Only one component, α-humulene (no. 25), was detected more often in JA-treated samples than in oviposition-induced ones. All other components of the headspace of JA-treated pine twigs were detected less frequently than in the volatile blend of oviposition-induced pine twigs.

**Table 1** List of volatiles collected from *P. sylvestris* twigs after different treatments. Only compounds that were identified in at least 5 out of 10 samples of oviposition-induced pine twigs are given. Values are given in ng/μl/sampling unit ± standard error. Number of replicates for each component are given within the columns of Figure 2.

| no. | compound | oviposition induced | control | JA induced | JA control |
|-----|----------|---------------------|---------|------------|------------|
| 1 | Thujene | 2 ± 0.4 | 1 ± 0.3 | 2 ± 0.3 | 2 ± 0.5 |
| 2 | α-Pinene | 46 ± 10.8 | 60 ± 14.4 | 100 ± 15.3 | 154 ± 34.5 |
| 3 | Camphene | 3 ± 0.8 | 5 ± 1.8 | 6 ± 1.0 | 5 ± 0.6 |
| 4 | Sabinene | 22 ± 5.3 | 21 ± 6.6 | 55 ± 15.7 | 43 ± 13.3 |
| 5 | β-Pinene | 29 ± 12.2 | 22 ± 4.4 | 79 ± 42.6 | 94 ± 31.7 |
| 6 | Myrcene | 170 ± 65.4 | 131 ± 27.6 | 142 ± 35.5 | 138 ± 35.0 |
| 7 | 3-Carene | 432 ± 76.3 | 423 ± 104.4 | 360 ± 130.5 | 359 ± 145.9 |
| 8 | 2-Carene | 4 ± 0.9 | 5 ± 1.9 | 10 ± 2.8 | 7 ± 2.5 |
| 9 | p-Cymene | 10 ± 1.6 | 12 ± 4.4 | 20 ± 7.1 | 16 ± 6.1 |
| 10 | Limonene[2] | 12 ± 2.3 | 13 ± 3.3 | 26 ± 7.8 | 18 ± 8.6 |
| 11 | β-Phellandrene[2] | 107 ± 37.0 | 70 ± 15.6 | 68 ± 34.1 | 65 ± 47.2 |
| 12 | (E)-β-Ocimene[1] | 2 ± 0.4 | 2 ± 0.6 | 7 ± 2.6 | 1 ± 0.6 |
| 13 | γ-Terpinene | 5 ± 1.0 | 7 ± 2.7 | 13 ± 3.3 | 9 ± 3.1 |
| 14 | unknown monoterpene hydrocarbon, B93, M136 | 3 ± 0.6 | 4 ± 1.9 | 8 ± 2.5 | 6 ± 2.2 |
| 15 | Terpinolene | 26 ± 5.7 | 36 ± 15.0 | 48 ± 18.1 | 34 ± 14.8 |
| 16 | p-Cymenene | 4 ± 0.8 | 5 ± 1.9 | 11 ± 4.1 | 9 ± 3.2 |
| 17 | 1,3,8-p-Menthatriene | 7 ± 1.5 | 10 ± 4.1 | 28 ± 9.6 | 21 ± 8.4 |
| 18 | unknown monoterpene, B119, M152 | 5 ± 1.1 | 8 ± 3.7 | 6 ± 2.3 | 4 ± 2.6 |
| 19 | (E,E)-2,6-Dimethyl-3,5, 7-octatrien-2-ol[1] | 7 ± 1.6 | 10 ± 5.2 | 8 ± 2.8 | 5 ± 3.4 |
| 20 | Thymyl methyl ether | 2 ± 0.7 | 1 ± 0.2 | 2 ± 0.6 | 2 ± 1.7 |
| 21 | Bornyl acetate | 1 ± 0.3 | 1 ± 0.5 | 1 ± 0.3 | 1 ± 0.3 |
| 22 | α–Copaene | 1 ± 0.5 | 1 ± 0.4 | 1 ± 0.3 | 1 ± 0.2 |
| 23 | β-Elemene | 2 ± 0.7 | 3 ± 1.0 | 3 ± 0.7 | 3 ± 1.3 |
| 24 | β-Caryophyllene | 15 ± 5.0 | 19 ± 5.2 | 19 ± 5.4 | 17 ± 6.3 |
| 25 | α-Humulene | 3 ± 0.7 | 3 ± 0.7 | 2 ± 0.7 | 2 ± 0.9 |
| 26 | (E)-β-Farnesene | 7 ± 2.7 | 5 ± 2.8 | 43 ± 19.8 | 12 ± 7.3 |
| 27 | Germacrene D | 6 ± 1.1 | 9 ± 2.0 | 3 ± 0.9 | 3 ± 1.0 |
| 28 | Bicyclogermacrene | 6 ± 1.8 | 7 ± 2.0 | 4 ± 1.1 | 3 ± 1.1 |
| 29 | α-Muurolene | 2 ± 0.6 | 2 ± 0.4 | 1 ± 0.2 | 1 ± 0.3 |
| 30 | γ-Cadinene | 7 ± 2.7 | 8 ± 2.5 | 4 ± 1.4 | 2 ± 0.9 |
| 31 | δ-Cadinene | 11 ± 3.8 | 10 ± 2.7 | 5 ± 1.4 | 3 ± 0.8 |

[1] For special information about quantities of (E)-β-ocimene and (E,E)-2,6-Dimethyl-3,5,7-octatrien-2-ol see chapter 6.

[2] Limonene and β-phellandrene were not separated from each other on DB-5 column. The equation given below was used to calculate the relative quantities of limonene and β-phellandrene. $X$: portion of limonene in TIC-peak of β-phellandrene + limonene, $Y$: portion of β-phellandrene in TIC-peak of β-phellandrene + limonene, $A_{68}$: peak area of ion trace 68; $A_{93}$: peak area of ion trace 93; $B_{68}$: portion of mass 68 of the total ion yield of a limonene pure sample [%]; $B_{93}$: portion of mass 93 of the total ion yield of a limonene pure sample [%]; $C_{93}$: portion of mass 93 of the total ion yield of a β-phellandrene pure sample [%].

$$X = \frac{A_{68} \cdot \dfrac{C_{93}}{B_{68}}}{A_{68} \cdot \dfrac{C_{93}}{B_{93}} + \left( A_{93} - A_{68} \cdot \dfrac{C_{93}}{B_{68}} \dfrac{B_{93}}{B_{68}} \right)} \qquad Y = 1 - X$$

Separation of enantiomeric monoterpenes revealed that α-pinene, camphene, β-pinene and limonene were present in both enantiomeric forms, whereas sabinene and β-phellandrene were found only as (−)-enantiomers in all samples. 3-carene was detected only as (+)-enantiomer. None of the treatments resulted in a significant change of the enantiomeric composition of the chiral monoterpenes analysed (Table 2).

Since neither oviposition nor JA-treatment induced a qualitative change of pine volatiles, samples were analysed statistically for quantitative differences. As characteristic for terpenoid contents in conifers (Latta et al. 2000, and references therein) the absolute quantities of emitted compounds varied much between the samples (Table 1). However, we wanted to elucidate whether quantities of volatiles from a test twig change relative to the respective control twig. Therefore, we calculated the relative amount of each volatile component emitted by a test- and its control twig based on the sum of the respective volatile component emitted by both twigs. Figure 2A shows that relative amounts of nearly all the components detected in the headspace of oviposition-induced pine twigs were about the same as those detected in the headspace of the respective control twigs. However, one sesquiterpene, (*E*)- β-farnesene (no.26), occurred in significantly higher relative quantities than in the controls, though the absolute amounts were not highly different (Table 1). Interestingly, relative amounts of (*E*)- β-farnesene were also significantly higher in JA-treated pine twigs when compared to the control (Figure 2B). Additionally, the headspace of JA-treated pine twigs contained α-muurolene (no. 29), γ-cadinene (no. 30), and δ-cadinene (no. 31) in significantly higher relative amounts than the controls. The relative amounts of several other components seem to be higher in JA-treated samples than in the controls, however, the number of samples in which these components were detected (N = 5 as defined minimum for statistical analysis) was too low for a valid statistical analysis.

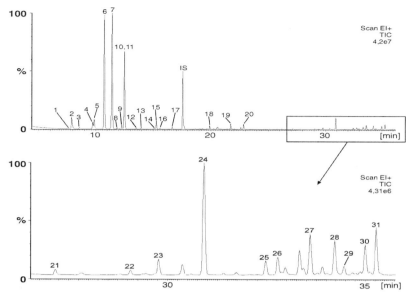

**Figure 1** Total ion current chromatogram of the headspace volatiles from pine twigs after egg deposition of *D. pini*. Numbers of peaks refer to compounds listed in Table 1. IS: internal standard (125 ng / µl methyl octanoate). Peaks without numbers denote compounds not present in more than five samples.

## Discussion

The headspace analyses of oviposition-induced pine twigs, JA-treated ones, and the respective controls detected numerous terpenoid components that are very typical constituents of conifer resin. Most components have been reported before from *Pinus sylvestris* (see Manninen et al. 2002 and references therein), with the exception of p-cymenene (no. 16), 1,3,8-p-menthatriene (no. 17), thymyl methyl ether (no. 20), (*E*)-β-farnesene (no. 26), and bicyclogermacrene (no. 28). Also (*F*,*E*)-2,6-dimethyl-3,5,7-octatrien-2-ol has not been identified in pine oleoresin before, but is known from *Ulmus minor* (Wegener et al. 2001), *Narcissus sp.* (van Dort et al. 1993) and several plant essential oils (e.g. Baser 2002). When compared to artificially damaged pine twigs, induction by sawfly egg deposition did not result in a qualitative change of the composition of the pine´s volatile blend, but only a significant relative increase in a single sesquiterpene, (*E*)-β-farnesene.

Since wounding of conifers is well-known to affect resin flow and formation of resin canals (e.g. Nebeker et al. 1995; Oven and Torelli 1999; Tomlin et al. 1998), also the mere artificial damage of pine needles mimicking the wounding conducted by a sawfly female prior to egg deposition was expected to cause an odour change when compared to undamaged pine. However, the artificially damaged pine here served as control since odour from these twigs did not attract the egg parasitoid (Hilker et al. 2002a). The impact of wounding and insect attack on the formation of oleoresin mono-, sesqui- and diterpenes of conifers has been investigated in several studies (see Gershenzon and Croteau 1991, Trapp and Croteau 2001, and references therein). With respect to the storage capacity of constitutively produced oleoresin, two wound responses of conifers may be differentiated (Gijzen et al. 1993):

(1)    Conifers like pines with well-developed resin ducts and high storage capacity for large amounts of constitutively produced oleoresin are known to translocate resin in response to damage to the site of wounding (e.g. Gijzen et al. 1993; and references therein). Nevertheless, increased activity of monoterpene cyclase induced by damage was found in needles of Ponderosa pine (*Pinus ponderosa*). Interestingly, feeding-induced enzyme activity was much higher than artificially wounding induced activity (Litvak and Monson 1998). However, when considering woody pine tissue, as for example of *Pinus taeda*, wound-induced resin flow is increased compared to constitutive flow (Lombadero et al. 2000), while no wound-inducibility of monoterpene synthase activity is detectable in stem tissue of this pine species (Phillips et al. 1999).

(2)    On the other hand, conifers like grand fir (*Abies grandis*) with low constitutive terpene biosynthetic activity show increased terpene biosynthesis in response to stem wounding due to the inducibility of mono-, sesqui- and diterpene synthases at the site of damage (Bohlmann et al. 1998; Steele et al. 1998, and references therein).

**Table 2** Enantiomeric composition [%] of chiral monoterpenes collected from differently treated pine twigs. Since a test and a control twig were always taken from the same branch, they were considered as paired samples. There were no significant differences between the ratios of enantiomers detected in test and control twigs (Wilcoxon matched pairs test, N = number of paired samples).

| component | oviposition | control | N | JA-induced | JA-control | N |
|---|---|---|---|---|---|---|
| (-)-α-Pinene | 51.5 | 49.3 | 7 | 31.1 | 30.6 | 10 |
| (+)-α-Pinene | 48.5 | 50.7 | | 68.9 | 69.4 | |
| (-)-Camphene | 95.8 | 93.4 | 6 | 81.2 | 85.6 | 6 |
| (+)-Camphene | 4.2 | 6.6 | | 18.8 | 14.4 | |
| (-)-Sabinene | 100.0 | 100.0 | 7 | 100.0 | 100.0 | 5 |
| (+)-Sabinene | 0.0 | 0.0 | | 0.0 | 0.0 | |
| (-)-β-Pinene | 100.0 | 98.5 | 7 | 99.2 | 98.3 | 9 |
| (+)-β-Pinene | 0.0 | 1.5 | | 0.8 | 1.7 | |
| (-)-3-Carene | 0.0 | 0.0 | 7 | 0.0 | 0.0 | 7 |
| (+)-3-Carene | 100.0 | 100.0 | | 100.0 | 100.0 | |
| (-)-Limonene | 95.8 | 95.1 | 5 | 92.0 | 96.7 | 8 |
| (+)-Limonene | 4.2 | 4.9 | | 8.0 | 3.3 | |
| (-)-β-Phellandrene | 100.0 | 100.0 | 7 | 100.0 | 100.0 | 8 |
| (+)-β-Phellandrene | 0.0 | 0.0 | | 0.0 | 0.0 | |

A comparison of the terpene patterns of wounded conifer tissue and of undamaged tissue revealed mainly quantitative differences rather than qualitative ones (Raffa and Smalley 1995; Popp et al. 1995). Particularly, woody tissue has been studied when analysing the effect of wounding on terpenoid synthesis in conifers (Gershenzon and Croteau 1991; Trapp and Croteau 2001, and references therein). For example, the relative amounts of monoterpene enantiomers isolated from trunk xylem of *Pinus caribaea* changed after insect attack, whereas no obvious change in relative amounts of the monoterpenes was detected in another pine species, *P. yunnanensis* (Fäldt et al. 2001). Similarly, defoliation did not induce a significant change of monoterpene concentrations in needles of several other pine species (Watt et al. 1991; Barnola et al. 1994) and clones of Douglas fir *Pseudotsuga menziesii* (Chen et al. 2002). When focusing on studies of needles from *P. sylvestris,* again no significant changes in the composition of monoterpenes, sesquiterpenes and resin acids were found after defoliation (Honkanen et al. 1999). Also Sadof and Grant (1997) did not find differences in monoterpene composition and ratio of enantiomers between volatile samples of wounded and unwounded *P. sylvestris*.

With respect to these known results from other studies and the ones obtained in our analyses, we conclude that neither wounding nor induction by egg deposition nor induction by JA leads to a qualitative change of the mono- and sesquiterpene pattern in *P. sylvestris*. Also no effect on the enantiomer composition has been detected so far in response to these treatments. When comparing the inducibility of *P. sylvestris* by egg deposition and JA, the latter treatment seems to have a stronger effect on the quantities of mono- and sesquiterpenes than oviposition (Figure 2). Thus, terpenoid-based responses of *P. sylvestris* were shown to be inducible by JA, as it is well-known for numerous angiosperms (Karban and Baldwin 1997; Boland et al. 1995, 1999; Gols et al. 1999; Rodriguez-Saona et al. 2001). Recently, Martin et al. (2002) and Fäldt et al. (2003) have shown that methyl jasmonate induces transient transcript accumulation of terpene synthases resulting in an increased terpenoid resin biosynthesis and terpenoid accumulation in *Picea abies*. Furthermore, other than terpenoid-based responses towards JA have been shown in gymnosperms (e.g. Kaukinen et al. 1996; Richard et al. 1999, 2000; Franceschi et al. 2002). In some angiosperm species, JA treatment results in only a quantitative shift within the volatile blend, while in others *de novo* volatile induction was shown (e.g. Hopke et al. 1994; Boland et al. 1995). In *P. sylvestris*, we could not detect any qualitative changes after JA-treatment, but only quantitative shifts in the volatile blend.

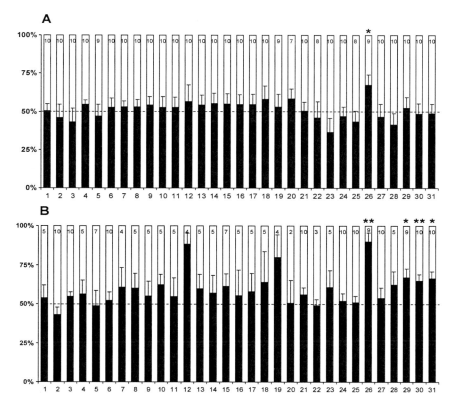

**Figure 2** Percentages of the relative amounts of compounds from treated twigs (dark columns) and control twigs (white columns) are given. The emitted amounts of a volatile component emitted by the test twigs and control twigs were considered 100%. **A**: Oviposition-induced twigs (dark columns) compared with artificially damaged twigs (white columns); **B**: Twigs treated with JA (dark columns) compared with untreated twigs (white columns). mean ± standard error, columns refer to the detected compounds given in Table 1, numbers inside the columns indicate number of paired samples in which the compounds have been identified. Statistical analysis: Wilcoxon matched pairs test, *: $p < 0.05$, **: $p < 0.01$.

Terpenoids (Langenheim 1994) seem to play a major role in the attraction of egg parasitoids to oviposition-induced plants. Not only the odour of oviposition-induced pine twigs, but also the volatile blend of elm leaves induced by egg deposition of the elm leaf beetle is dominated by terpenoids (Wegener et al. 2001). The odour of JA-treated elm leaves which also attracts the elm leaf beetle's egg parasitoid was shown to consist almost exclusively of terpenoids. However, $(E)$-β-farnesene was not detected in oviposition-induced and JA-treated elm leaves (Wegener et al. 2001). In contrast to oviposition-induced pine, elm leaves induced by egg deposition showed both quantitative and qualitative changes of their volatile blends.

($E$)-β-Farnesene, the only component that was emitted in significantly higher amounts in both oviposition-induced pine twigs and JA-treated ones, is a common terpenoid component constitutively present in several plants (Teuscher and Lindequist 1994). Numerous studies have shown increased amounts or *de novo* production of this sesquiterpene in plants induced by feeding herbivores (e.g. Turlings et al. 1991; 1993; Takabayashi et al. 1995; Bolter et al. 1997; Paré and Tumlinson 1997; De Moraes et al. 1998; Gols et al. 1999; Paré et al. 1999), by mechanical damage (e.g. McAuslane and Alborn 1998), and by treatment with regurgitate (Turlings et al. 1993), volicitin or JA (e.g. Rodriguez-Saona et al. 2001; Schmelz et al. 2001). The increased amount of ($E$)-β-farnesene emitted is not restricted to the leaves under attack, but also adjacent, but non-damaged leaves of a plant under attack by feeding herbivores may release higher amounts of this sesquiterpene (e.g. Röse et al. 1998). Thus, both systemically oviposition-induced and feeding-induced plants may respond by increased emission of ($E$)-β-farnesene.

Numerous insect species are able to respond to ($E$)-β-farnesene. For example, this sesquiterpene is well-known as alarm pheromone in aphids (Nault et al. 1973). Both herbivorous insects (e.g. Bengtsson et al. 2001) and predatory insects show EAG responses towards ($E$)-β-farnesene (e.g. Zhu et al. 1999; Al Abassi et al. 2000; Weissbecker et al. 2000). Also aphid parasitoids have been shown to respond electrophysiologically as well as behaviourally towards ($E$)-β-farnesene (Micha and Wyss 1996; Du et al. 1998). To elucidate the role of ($E$)-β-farnesene within the tritrophic system of *P. sylvestris,* the sawfly *D. pini,* and its egg parasitoid *C. ruforum,* it will be necessary to study the behavioural and electrophysiological responses of the egg parasitoid towards synthetic ($E$)-β-farnesene. Dicke and Vet (1999) discussed the informational value of a herbivore-induced quantitative change in the volatile blend of a plant for carnivores. Up to now, we do not know how carnivores are attracted towards a quantitatively changed, herbivore-induced plant odour blend. Future studies need to investigate whether the carnivores rely on a specific plant volatile that is released in specific quantities or whether they respond finely tuned to the pattern of ratios of the volatiles within a blend emitted by a herbivore-induced plant.

## *Acknowledgements*

Many thanks are due to Frank Müller for his assistance in volatile collections and GC-MS-analyses. We are also very grateful to Ute Braun who helped to rear *Diprion pini*. We also thank two anonymous reviewers for constructive comments on the manuscript. Stefan Schulz likes to thank the Fonds der Chemischen Industrie for financial support. This study was supported by the Deutsche Forschungsgemeinschaft (DFG Hi 416/11-1,2).

## References

Adams, R. P. 1995. Identification of essential oil - components by gas chromatography / mass spectroscopy. Allured Publishing Corporation, Carol Stream, Illinois USA.

Aducci, P. 1997. Signal Transduction in Plants. Molecular and Cell Biology Updates. Birkhäuser, Basel.

Agrawal, A. A., Tuzun, S., Bent, E. 1999. Induced Plant Defenses against Pathogens and Herbivores. Biochemistry, Ecology, and Agriculture. APS Press, St. Paul, Minnesota.

Al Abassi, S., Birkett, M. A., Petterson, J., Pickett, J. A., Wadhams, L. J., Woodcock, C. M. 2000. Response of the seven-spot ladybird to an aphid alarm pheromone and an alarm pheromone inhibitor mediated by paired olfactory cells. *J. Chem. Ecol.* 26:1765-1771.

Baldwin, I. T. 1994. Chemical changes rapidly induced by folivory. pp. 1- 23, *in* Bernays E.A. (ed.) Insect-Plant Interactions. Vol. 5. CRC Press, Boca Raton.

Barnola, L. F., Hasegawa, M., Cedono, A. 1994. Mono- and sesquiterpene variation in *Pinus caribaea* needles and its relationship to *Atta laevigata* herbivory. *Biochem. Syst. Ecol.* 22:437-445.

Baser, K. H. C., Demirci, B. Kirimer, N. 2002. Compositions of the essential oils of four *Helichrysum* species from Madagascar. *J. Essent. Oil Res.* 14:53-55.

Beale, M. H, Ward, J. L. 1998. Jasmonates: key players in plant defence. *Nat. Prod. Rep.* 1998:533-547.

Bengtsson, M., Backman, A. C., Liblikas, I., Ramirez, M. I., Borg-Karlson, A. K., Ansebo, L., Anderson, P., Loefqvist, J., Witzgall, P. 2001. Plant odor analysis of apple: Antennal response of codling moth females to apple volatiles during phenological development. *J. Agric. Food Chem.* 49:3736-3731.

Bohlmann, J., Crock, J., Jetter, R., Croteau, R. 1998. Terpenoid-based defences in conifers: cDNA cloning, characterization, and functional expression of wound-inducible (E)-β-bisabolene synthase from grand fir (*Abies grandis*). *Proc. Natl. Acad. Sci. USA* 95:6756-6761.

Boland, W., Koch, T., Krumm, T., Piel, J., Jux, A. 1999. Induced biosynthesis of insect semiochemicals in plants, pp. 110-126, *in* Chadwick, D. J., Goode, J. A. (eds.). Insect-Plant Interactions and Induced Plant Defence. John Wiley & Sons Ltd, Chicester.

Boland, W., Hopke, J., Donath, J., Nüske, J., Bublitz, F. 1995. Jasmonic acid and coronatine induce odor production in plants. *Angew. Chem. Int. Ed. Engl.* 34:1600-1602.

Bombosch, S., Ramakers, P. M. J. 1976. Zur Dauerzucht von *Gilpinia hercyniae* Htg. *Z. Pflanzenkrank. Pflanzenschutz* 83:40-44.

Bolter, C. J., Dicke, M., van Loon, J. J. A., Visser, J. H., Posthumus, M. A. 1997. Attraction of Colorado potato beetle to herbivore-damaged plants during herbivory and after its termination. *J. Chem. Ecol.* 23:1003-1023.

Chen, Z., Kolb, T. E., and Clancy, K. M. 2002. The role of monoterpenes in resistance of Douglas fir to western spruce budworm defoliation. *J. Chem. Ecol.* 28:897-920.

Creelman, R. A., Mullet, J. E. 1997. Biosynthesis and action of jasmonates in plants. *Annu. Rev. Plant Physiol. Plant Mol. Biol.* 48:355-381.

De Moraes, C. M., Lewis, W. J., Paré, P. W., Alborn, H. T., Tumlinson, J. H. 1998. Herbivore-infested plants selectively attract parasitoids. *Nature* 393:570-573.

Dicke, M. 1999. Evolution of induced indirect defence of plants, pp. 62-88, *in*: Tollrian R. and Harvell, C.D. (eds.) The Ecology and Evolution of Inducible Defenses. Princeton University Press, Princeton, NJ.

Dicke, M. 1994. Local and systemic production of volatile herbivore-induced terpenoids: their role in plant-carnivore mutualism. *J. Plant Physiol.* 143:465-472.

Dicke, M., van Loon, J. J. A. 2000. Multitrophic effects of herbivore-induced plant volatiles in an evolutionary context. *Entomol. Exp. Appl.* 97:237-249.

Dicke, M., Vet, L. E. M. 1999. Plant-carnivore interactions: evolutionary and ecological consequences for plant, herbivore and carnivore, pp. 483-520, *in* Olff, H., Brown, V. K., and Drent, R. H. (eds.). Herbivores Between Plants and Predators. Blackwell Science Oxford.

Dicke, M., Gols, R., Ludeking, D., Posthumus, M.A. 1999. Jasmonic acid and herbivory differentially induce carnivore-attracting plant volatiles in lima bean plants. *J. Chem. Ecol.* 25:1907-1922.

Dicke, M., Van Beek, T. A., Posthumus, M. A., Ben Dom, N., Van Bokhoven, H., De Groot, A. E. 1990. Isolation and identification of volatile kairomone that affects acarine predator-prey interactions. *J. Chem. Ecol.* 16:381-396.

Du, Y., Poppy, G. M., Powell, W., Pickett, J. A., Wadhams, L. J., and Woodcock, C. M. 1998. Identification of semiochemicals released during aphid feeding that attract parasitoid *Aphidius ervi*. *J. Chem. Ecol.* 24:1355-1369.

Duffey, S. S., Stout, M. J. 1996. Antinutritive and toxic components of plant defense against insects. *Arch. Insect Biochem. Physiol.* 32:3-37.

Edwards, P. J., Wratten, S. D. 1987. Ecological significance of wound induced changes in plant chemistry, pp. 213-219, *in* Labeyrie, V., Fabres, G., and Lachaise, D. (eds.). Insects - Plants: Proc. 6th Int. Symp. Insect – Plant Relationships. Dr. W. Junk Publ., The Hague.

Eichhorn, O. 1976. Dauerzucht von *Diprion pini* L. (Hym.: Diprionidae) im Laboratorium unter Berücksichtigung der Fotoperiode. *Anz. Schädlingskde., Pflanzenschutz, Umweltschutz* 49:38-41.

Fäldt, J., Martin, D., Miller, B., Rawat, S., Bohlmann, J. 2003. Traumatic resin defense in Norway spruce (*Picea abies*): Methyl jasmonate-induced terpene synthase gene expression, and cDNA cloning and functional characterization of (+)-3-carene synthase. *Plant Mol. Biol.* 51:119-133.

Fäldt, J., Sjödin, K., Persson, M., Valterova, I., Borg-Karlsson, A. K. 2001. Correlations between selected monoterpene hydrocarbons in the xylem of six *Pinus* (Pinaceae) species. *Chemoecology* 11:97-106.

Franceschi, V. R., Krekling, T., Chrisitansen, E. 2002. Application of methyl jasmonate on *Picea abies* (Pinaceae) stems induces defense-related responses in phloem and xylem. *Am. J. Bot.* 89:578-586

Fugmann, B., Lang-Fugmann, S., Steglich, W. 1997. Römpp-Lexikon Naturstoffe. Georg Thieme, Stuttgart.

Gershenzon, J., Croteau, R. 1991. Terpenoids. pp. 165-219, *in* Rosenthal, G. A. and Berenbaum, M. R. (eds.). Herbivores. Their Interactions with Secondary Plant Metabolites. Vol. 1. The Chemical Participants. Academic Press, New York.

Gijzen, M., Lewinsohn, E., Savage, T. J., Croteau, R. B. 1993. Conifer Monoterpenes. pp. 8-22, *in* Teranishi, R., Buttery, R. G., and Sugisawa, H. (eds.). Bioactive Volatile Compounds from Plants. ACS Symposium Series 525, Washington.

Gols, R., Posthumus, M. A., Dicke, M. 1999. Jasmonic acid induces the production of gerbera volatiles that attract the biological control agent *Phytoseiulus persimilis*. *Entomol. Exp. Appl.* 93:77-86.

Hilker, M., Meiners, T. 2002. Induction of plant responses towards oviposition and feeding of herbivorous arthropods: a comparison. *Entomol. Exp. Appl.* 104:181-192.

Hilker, M., Kobs, C., Varama, M., Schrank, K. 2002a. Insect egg deposition induces *Pinus* to attract egg parasitoids. *J. Exp. Biol.* 205:455-461.

Hilker, M., Rohfritsch, O., Meiners, T. 2002b. The plant´s response towards insect egg deposition. pp. 205-234, *in* Hilker, M. and Meiners, T. (eds.). Chemoecology of Insect Eggs and Egg Deposition. Blackwell Publishing, Berlin.

Honkanen, T., Haukioja, E., Kitunen, V. 1999. Responses of *Pinus sylvestris* branches to simulated herbivory are modified by tree sink/source dynamics and by external resources. *Funct. Ecol.* 13:126-140.

Hopke, J., Donath, J., Blechert, S., Boland, W. 1994. Herbivore-induced volatiles: the emission of acyclic homoterpenes from leaves of *Phaseolus lunatus* and *Zea mays* can be triggered by a β-glucosidase and jasmonic acid. *FEBS Letters* 352:146-150.

Joulain, D., König, W. A. 1998. The Atlas of Spectral Data of Sesquiterpene Hydrocarbons. E.-B. Verlag, Hamburg..

Karban, R., Baldwin, I. T. 1997. Induced Responses to Herbivory. The University Press of Chicago, Chicago.

Kaukinen, K. H., Tranbarger, T. J., Misra, S. 1996. Post germination induced and hormonally dependent expression of low molecular weight heat shock protein genes in Douglas fir. *Plant Mol. Biol.* 30:1115-1128.

Kessler, A., Baldwin, I. T. 2002. Plant responses to insect herbivory: the emerging molecular analysis. *Annu. Rev. Plant Biol.* 53:299-328.

König, W. A., Krüger, A., Icheln, D., Runge, T. 1992. Enantiomeric composition of the chiral constituents in essential oils. *J. High Resol. Chromatogr.* 15:184-189.

Koch, T., Krumm, T., Jung, V., Engelberth, J., Boland, W. 1999. Differential induction of plant volatile biosynthesis in the lima bean by early and late intermediates of the octadecanoid-signaling pathway. *Plant Physiol.* 121:153-162.

Langenheim, J. H. 1994. Higher plant terpenoids: A phytocentric overview of their ecological roles. *J. Chem. Ecol.* 20:1223-1280.

Latta, R. G., Linhart, Y. B., Lundquist, L., Snyder, M.A. 2000. Patterns of monoterpene variation within individual trees in ponderosa pine. *J. Chem. Ecol.* 26:1341-1357.

Litvak, M. E., Monson, R. K. 1998. Patterns of induced and constitutive monoterpene production in conifer needles in relation to insect herbivory. *Oecologia* 114:531-540.

Lombardero, M. J., Ayres, M. P., Lorio Jr., P. L., Ruel, J. J. 2000. Environmental effects on constitutive and inducible resin defences in *Pinus taeda*. *Ecol. Lett.* 3:329-339.

Manninen, A. M., Tarhanen, S., Vuorinen, M., Kainulainen, P. 2002. Comparing the variation of needle and wood terpenoids in Scots pine provenances. *J. Chem. Ecol.* 28:211-228.

Martin, D., Tholl, D., Gershenzon, J., Bohlmann, J. 2002. Methyl jasmonate induces traumatic resin ducts, terpenoid resins biosynthesis, and terpenoid accumulation in developing xylem of Norway Spruce stems. *Plant Physiol.* 129:1003-1018.

McAuslane, H. J., Alborn, H. T. 1998. Systemic induction of allelochemicals in glanded and glandless isogenic cotton by *Spodoptera exigua* feeding. *J. Chem. Ecol.* 24:399-416.

Meiners, T., Hilker, M. 2000. Induction of plant synomones by oviposition of a phytophagous insect. *J. Chem. Ecol.* 26:221-232.

Meiners, T., Hilker, M. 1997. Host location in *Oomyzus gallerucae* (Hymenoptera: Eulophidae), an egg parasitoid of the elm leaf beetle *Xanthogaluruca luteola* (Coleoptera: Chrysomelidae). *Oecologia* 112:87-93.

Micha, S. G., Wyss, U. 1996. Aphid alarm pheromone (E)-$\beta$-farnesene: a host finding kairomone for the aphid primary parasitoid *Aphidius uzbekistanicus* (Hymenoptera: Aphidiinae). *Chemoecology* 7:132-139.

Moore, G. E., Clark, E. W. 1968. Suppressing microorganisms and maintaining turgidity in coniferous foliage used to rear insects in the laboratory. *J. Econ. Entomol.* 61:1030-1031.

Nault, L. R., Edwards, L. J., Styer, W.E. 1973. Aphid alarm pheromones: Secretion and reception. *Environ. Entomol.* 2:101-105.

Nebeker, T. E., Schmitz, R. F., Tisdale, R. A. 1995. Comparison of oleoresin flow in relation to wound size, growth rates, and disease status of lodgepole pine. *Can. J. Bot.* 73:370-375.

Oven, P., Torelli, N. 1999. Response of the cambial zone in conifers to wounding. *Phyton* 39:133-137.

Ozawa, R., Arimura, G., Takabayashi, J., Shimoda, T., Nishioka, T. 2000. Involvement of jasmonate- and salicylate-related signaling pathways for the production of specific herbivore-induced volatiles in plants. *Plant Cell Physiol.* 41:391-398.

Paré, P. W., Tumlinson, J. H. 1997. *De novo* biosynthesis of volatiles induced by insect herbivory in cotton plants. *Plant Physiol.* 114:1161-1167.

Paré, P. W., Lewis, W. J., Tumlinson, J. H. 1999. Induced plant volatiles: Biochemistry and effects on parasitoids. pp. 167-180, *in* Agrawal, A. A., Tuzun, S., and Bent, E. (eds.). Induced Plant Defenses against Pathogenes and Herbivores. APS Press, St. Paul.

Petrakis, P. V., Tsitsimpikou, C., Tzakou, O., Couladis, M., Vagias, C., Roussis, V. 2001. Needle volatiles from *Pinus* species growing in Greece. *Flavour Fragr. J.* 16:249-252.

Phillips, M. A., Savage, T. J., Croteau, R. 1999. Monoterpene synthases of loblolly pine (*Pinus taeda*) produce pinene isomers and enantiomers. *Arch. Biochem. Biophys.* 372:197-204.

Popp, M. P., Johnson, J. D., Lesney, M. S. 1995. Characterization of the induced response of slash pine to inoculation with bark beetle vectored fungus. *Tree Physiol.* 15:619-623.

Price, P.W. 1986. Ecological aspects of host plant resistance and biological control: Interactions among three trophic levels. pp. 11-30, *in* Boethel, D. J. and Eikenbary, R. D. (eds.). Interactions of Plant Resistance and Parasitoids and Predators of Insects. Ellis Hoerwood, Chichester.

Price, P. W., Bouton, C. E., Gross, P., McPheron, B. A., Thompson, J. N., Weis, A. E. 1980. Interactions among three trophic levels: Influence of plants on interactions between insect herbivores and natural enemies. *Annu. Rev. Ecol. Syst.* 11:41-65.

Raffa, K. F., Smalley, E. B. 1995. Interaction of pre-attack and induced monoterpene concentrations in host conifer defense against bark beetle-fungal complexes. *Oecologia* 102:285-295.

Richard, S., Lapointe, G., Rutledge, R. G., Seguin, A. 2000. Induction of chalcone synthase in white spruce by wounding and jasmonate. *Plant Cell Physiol.* 41:982-987.

Richard, S., Drevet, C., Jouanin, L., Sequin, A. 1999. Isolation and characterization of a cDNA clone encoding a putative white spruce glycine-rich RNA binding protein. *Gene* 240:379-388.

Rodriguez-Saona, C., Crafts-Brandner, S. J., Paré, P. W., Henneberry, T. J. 2001. Exogenous methyl jasmonate induces volatile emissions in cotton plants. *J. Chem. Ecol.* 27:679-695.

Röse, U. S. R., Lewis, W. J., Tumlinson, J. H. 1998. Specificity of systemically released cotton volatiles as attractants for specialist and generalist parasitic wasps. *J. Chem. Ecol.* 24:303-319.

Sadof, C. S., Grant, G. G. 1997. Monoterpene composition of *Pinus sylvestris* varieties resistant and susceptible to *Dioryctria zimmermani*. *J. Chem. Ecol.* 23:1917-1927.

Schmelz, E. A., Alborn, H. T., Tumlinson, J. H. 2001. The influence of intact-plant and excised-leaf bioassay designs on volicitin- and jasmonic acid-induced sesquiterpene volatile release in *Zea mays*. *Planta* 214:171-179.

Sembdner, G., Parthier, B. 1993. The biochemistry and the physiological and molecular actions of jasmonates. *Annu. Rev. Plant Physiol. Plant Mol. Biol.* 44:569-589.

Staswick,P. E., Lehman, C. C. 1999. Jasmonic acid-signaled responses in plants. pp. 117-136, *in* Agrawal, A. A., Tuzun, S., and Bent, E. (eds.). Induced Plant Defenses against Pathogens and Herbivores. APS Press, St. Paul, Minnesota.

Steele, C. L., Katoh, S., Bohlmann, J., Croteau, R. 1998. Regulation of oleoresinosis in grand fir (*Abies grandis*). *Plant Physiol.* 116:1497-1504.

Stout, M. J., Bostock, R. M. 1999. Specificity of induced responses to arthropods and pathogens. pp. 183-211, *in* Agrawal, A. A, Tuzun, S., and Bent, E. (eds.). Induced Defenses Against Pathogens and Herbivores. APS Press, St. Paul, Minnesota.

Takabayashi, J., Takahashi, S., Dicke, M., Posthumus, M.A. 1995. Developmental stage of herbivore *Pseudaletia separata* affects production of herbivore-induced synomone by corn plants. *J. Chem. Ecol.* 21:273-287.

Takabayashi, J., Dicke, M., Posthumus, M. A. 1994. Volatile herbivore-induced terpenoids in plant-mite interactions: Variation caused by biotic and abiotic factors. *J. Chem. Ecol.* 20:1329-1354.

Teuscher, E., Lindequist, U. 1994. *Biogene Gifte*. Gustav Fischer, Stuttgart.

Thaler, J. S. 1999. Jasmonate-inducible plant defences cause increased parasitism of herbivores. *Nature* 399:686-688.

Tomlin, E. S., Alfaro, R. I., Borden, J. H., He, F. 1998. Histological response of resistant and susceptible white spruce to simulated white pine weevil damage. *Tree Physiol.* 18:21-28.

Trapp, S., Croteau, R. 2001. Defensive resin biosynthesis in conifers. *Annu. Rev. Plant Physiol. Plant Mol. Biol.* 52:689-724.

Turlings, T. C. J., McCall, P. J., Alborn, H. T., Tumlinson, J. H. 1993. An elicitor in caterpillar oral secretions that induces corn seedlings to emit chemical signals attractive to parasitic wasps. *J. Chem. Ecol.* 19:411-425.

Turlings, T. C. J., Tumlinson, J. H., Heath, R. R., Proveaux, A. T., Doolittle, R. E. 1991. Isolation and identification of allelochemicals that attract the larval parasitoid, *Cotesia marginiventris* (Cresson), to the microhabitat of one of its hosts. *J. Chem. Ecol.* 17:2235-2251.

Turlings, T. C. J., Tumlinson, J. H., Lewis, W. J. 1990. Exploitation of herbivore-induced plant odors by host-seeking parasitic wasps. *Science* 250:1251-1253.

van den Dool, J., Kratz, P. D. 1963. A generalization of the retention index system including linear programmed gas-liquid partition chromatography. *J. Chromatogr.* 11:463.

van Dort, H. M., Jagers, P. P., ter Heide, R., van der Weerdt, A. J. A. 1993. *Narcissus trevithian* and *Narcissus geranium*: Analysis and synthesis of compounds. *J. Agric. Food Chem.* 41:2063-2075.

Walling, L. L. 2000. The myriad plant responses to herbivores. *J. Plant Growth Regul.* 19:195-216.

Watt, A. D., Leather, S. R., Forrest, G. I. 1991. The effect of previous defoliation of pole-stage lodgepole pine on plant chemistry, and on the growth and survival of pine beauty moth (*Panolis flammea*) larvae. *Oecologia* 86:31-35.

Wegener, R., Schulz, S., Meiners, T., Hadwich, K., Hilker, M. 2001. Analysis of volatiles induced by oviposition of elm leaf beetle *Xanthogaleruca luteola* on *Ulmus minor. J. Chem. Ecol.* 27:499-515.

Weissbecker, B., van Loon, J. J. A., Posthumus, M. A., Bouwmeester, H. J., Dicke, M. 2000. Identification of volatile potato sesquiterpenoids and their olfactory detection by the two-spotted stinkbug *Perillus bioculatus. J. Chem. Ecol.* 26:1433-1445.

Zhu, J., Cossé, A. A., Obrycki, J. J., Boo, K. S., Baker, T.C. 1999. Olfactory reactions of the twelve-spotted lady beetle, *Coleomegilla maculata* and the green lacewing, *Chrysoperla carnea*, to semiochemicals released from their prey and host plant: electroantennogram and behavioral responses. *J. Chem. Ecol.* 25:1163-1177.

# Chapter 4

# How Does an Egg Parasitoid Recognize Odour
# of Pine with Host Eggs ?

**Abstract.** Pine (*Pinus sylvestris*) has been shown to produce a volatile bouquet in response to oviposition of the herbivorous sawfly *Diprion pini* that attracts egg parasitoids killing the eggs. This plant response has been interpreted as a preventive defensive strategy against herbivory. Previous analyses of headspace volatiles of oviposition-induced pine twigs revealed no qualitative, but quantitative changes compared to respective controls. Especially the sesquiterpene $(E)$-β-farnesene was emitted in significantly higher amounts by oviposition-induced pine twigs. In this study we investigated whether $(E)$-β-farnesene attracted the egg parasitoid *Chrysonotomyia ruforum*, which is specialized on diprionid eggs. In olfactometer bioassays we tested the behavioural response of experienced *C. ruforum* females to different concentrations of synthetic $(E)$-β-farnesene (0.01, 0.1 and 1μg/μl hexane). The egg parasitoid did not respond to this sesquiterpene at neither concentration tested. However, egg parasitoids responded significantly to $(E)$-β-farnesene when this compound was offered in combination with the volatile blend emitted from a pine twig carrying no eggs. This response was dependent on the applied concentration of $(E)$-β-farnesene. When a concentration of 0.01μg/μl $(E)$-β-farnesene was offered on the background odour of an egg-free pine twig, no significant response of the egg parasitoid was detected, whereas a concentration of 0.1μg/μl attracted the parasitoids significantly. In contrast, a 10-fold higher concentration of $(E)$-β-farnesene (1μg/μl) elicited a significant avoidance behaviour of the parasitoids. Further bioassays with other components $((E)$-β-caryophyllene, δ-cadinene) of the odour blend of pine were conducted in combination with the volatile blend from egg-free pine as background odour to test whether the detected response of egg parasitoids to $(E)$-β-farnesene was specific for this sesquiterpene. None of these compounds tested on the background of odour from an egg-free pine twig were attractive to the egg parasitoid. These results suggest that the egg parasitoids responded specifically to $(E)$-β-farnesene, however, only when this compound was experienced in the "right" context, i.e. with a background odour of non-oviposition-induced pine volatiles.

**Keywords.** *Diprion pini*, $(E)$-β-farnesene, egg deposition, egg parasitoid, induced defense, *Pinus sylvestris*, sesquiterpene.

**Introduction**

Carnivorous arthropods are known to use volatile cues that are released by plants infested with herbivorous insects (e.g. Dicke and van Loon 2000, Hilker and Meiners 2002; Hilker et al. 2002b; Turlings et al. 2002). Plants emit a complex blend of compounds which in response to herbivory may be induced to be altered either qualitatively or quantitatively (e.g. Dicke 1999). Qualitative alterations comprise the production of novel compounds that are not emitted by uninfested plants. On the other hand, plants may also respond to herbivore infestation by emission of a volatile pattern that is qualitatively similar to the blend emitted by intact or mechanically damaged plants. In this case, the emission rate is much higher or the quantitative composition of the blend is changed (Dicke and van Poecke 2002). Though herbivore-induced plant volatiles can provide carnivores like parasitoids with specific information about the attacking herbivore, the composition of plant volatile blends can also be highly variable (overview given by Dicke 1999). The high geno- and phenotypic variability of herbivore-induced plant volatiles might limit the reliability of these cues for parasitoids (Dicke and Vet 1999; Dicke and Hilker 2003). Many parasitoids have the ability to learn plant odours associatively during a host encounter, thus enabling them to adjust their responses to varying host-related cues (Turlings et al. 1993; Vet et al. 1995).

Regarding the high variability of plant volatiles the question arises which differences between herbivore-induced and non-induced volatile blends do parasitoids learn to discriminate the "right" from the "wrong" blend. In this study, we investigated this question for an egg parasitoid that learns to respond to plant volatiles induced by the egg deposition of its herbivorous host. Egg deposition of the herbivorous sawfly *Diprion pini* L. (Hymenoptera, Diprionidae) induced Scots pine (*Pinus sylvestris* L.) to emit volatiles that attract the specialized egg parasitoid *Chrysonotomyia ruforum* Krausse (Hymenoptera, Eulophidae) (Hilker et al. 2002a). Egg parasitoids were not attracted to volatiles from artificially wounded pines. The artificial wounding mimicked the damage of the pine needles inflicted by sawfly females with the sclerotized ovipositor valves prior to egg deposition. Moreover, the application of jasmonic acid (JA), a phytohormone widely involved in herbivore-induced defence responses in plants (reviewed by de Bruxelles and Roberts 2001; Gatehouse 2002; Schaller and Weiler 2002), induced pines to emit volatiles that attracted females of *C. ruforum* (Hilker et al. 2002a).

In order to detect the pine volatiles induced by egg deposition of *D. pini,* we analysed the volatiles of the headspace of oviposition-induced pine twigs and compared them to volatiles from artificially wounded (non-attractive) pine twigs. Additionally, odour from JA-treated pine twigs was compared to odour from untreated and undamaged pine twigs. Neither egg deposition nor JA-treatment induced a qualitative change in the volatile blend of pine compared to the respective controls (Mumm et al. 2003). Oviposition-induced and JA-treated pine twigs emitted the same compounds, i.e. mainly mono- and sesquiterpenes, as the respective controls. Except of one component, no significant quantitative changes were detected when comparing oviposition-induced twigs and the respective controls. However, the sesquiterpene (*E*)-β-farnesene was emitted in significantly higher amounts by oviposition-induced pine twigs compared to the controls (Mumm et al. 2003). Interestingly, among three other sesquiterpenes (α-muurolene, γ-cadinene and δ-cadinene), (*E*)-β-farnesene was also emitted in significantly higher amounts by pine twigs treated with JA compared to untreated controls. Thus, (*E*)-β-farnesene was the only component that was released in significantly larger amounts from pine twigs induced by egg deposition and those treated with jasmonic acid (Mumm et al. 2003).

In this study, we investigated whether (*E*)-β-farnesene is utilized by *C. ruforum* as chemical cue to locate pine infested with eggs of *D. pini.* The behavioural response of *C. ruforum* to different concentrations of (*E*)-β-farnesene was tested in an olfactometer. Background odour is known to affect the response of insects to single volatile components (Kelling et al. 2002, Smith 1998, and references therein). Therefore, we hypothesized that *C. ruforum* shows a different behavioural response to (*E*)-β-farnesene if this is offered in combination with the natural volatile blend of the host plant.

In order to test whether the egg parasitoid's behavioural response which we detected for (*E*)-β-farnesene was specific for this sesquiterpene, also the parasitoid's response to two further terpenoid components was studied. (*E*)-β-Caryophyllene was chosen since it was identified as the predominating sesquiterpene in the headspace of attractive oviposition-induced and JA-treated pine (Mumm et al. 2003). However, neither egg deposition nor treatment with jasmonic acid resulted in an increased emission of (*E*)-β-caryophyllene compared to the respective controls (Mumm et al. 2003). Furthermore, δ-cadinene was selected as test component since it was emitted in significantly higher amounts by the attractive twigs treated with jasmonic acid when compared to untreated controls. However, oviposition-induced twigs did not release higher amounts of this terpene (Mumm et al. 2003). (*E*)-β-Caryophyllene and δ-cadinene were both tested on the background of volatiles from pine twigs without eggs.

**Materials and methods**

*Plants and insects*

Branches of *Pinus sylvestris* L. used for experiments and rearing were detached from crowns of 15- to 35-year-old trees in the forests near Berlin. All stems were cleaned and sterilized according to the method of Moore and Clark (1968). *Diprion pini* L. was reared continuously in the laboratory on cut pine twigs as described by Bombosch and Ramakers (1976) and Eichhorn (1976) at 25±1°C, L/D 18:6 h, and 70 % relative humidity. The egg parasitoid *Chrysonotomyia ruforum* Krausse was obtained from parasitized eggs of *D. pini* and *Neodiprion sertifer* Geoffroy collected in the field in France (near Fontainebleau) and central and southern Finland. Parasitized eggs were kept in petri dishes (i.d. 9 cm) in a climate chamber at 10 °C. To induce parasitoid emergence, needles with parasitized eggs were placed in a climate chamber at 25 °C, 18L:6D photoperiod and 70 % relative humidity. Emerging adults were collected daily and transferred in small perspex tubes (75mm long, 15mm i.d.) covered with gauze at one end. A cotton-wool plug moistened with an aqueous honey solution closed the other end. The parasitoids were kept at 10 °C, 18L:6D, until they were used for bioassays.

Parasitoid females used for the bioassays were experienced with the plant-host complex as described by Hilker et al. (2002a). Two days prior to the bioassays, parasitoids were given contact with male parasitoids and a plant-host complex, consisting of a pine twig carrying eggs of *D. pini*, adult sawflies and a cotton-wool pad with aqueous honey solution. After 24 h, parasitoids were removed from the plant-host complex and kept in clean perspex tubes provided with the aqueous honey solution only for another 24 h prior to the experiments.

*Olfactometer bioassay – general procedures and data collection*

All bioassays were conducted in a four-arm olfactometer (Pettersson 1970; Vet et al. 1983) as described by Hilker et al. (2002a) in detail. The airflow was adjusted to 155ml min[-1]. When starting a bioassay, a parasitoid female was introduced into the arena of the olfactometer. We recorded how long the parasitoid spent walking within each of the four odour fields over a period of 600 s using the Observer program 3.0 (Noldus, Wageningen, The Netherlands). Data obtained from parasitoids that walked less than 300 s were discarded.

*Chemicals (Terpenes)*

(*E*)-β-Farnesene was synthesized according to a modified procedure of Kang et al. (1987). To a stirred solution of farnesyl chloride (0.25 g, 1.0 mmol, Aldrich, Steilheim, Germany), 18-crown-8-ether (0.53g, 2.0 mmol, Aldrich, Steilheim, Germany) was added followed by KOtBu (1.1 g 10 mmol, Aldrich, Steilheim, Germany) in dry THF (5ml, Aldrich, Steilheim, Germany). The mixture was stirred at 40 °C for 24 h. 30 ml t-Butylmethylether was added to the reaction mixture and washed with water and saturated NaCl solution. The mixture was dried over $Na_2SO_4$ and then concentrated *in vacuo*. The crude product was purified and separated by solid phase extraction with silica gel as stationary phase. (*E*)-β-Farnesene was eluted with n-hexane : ethyl acetate (9.5 : 0.5, v / v). With this method, (*E*)-β-farnesene of 99% purity was obtained. The other terpenes were purchased as reference compounds from commercial companies: (*E*)-β-caryophyllene (80%; Aldrich, Steilheim, Germany), and δ-cadinene (>97%; Fluka, Buchs, Switzerland).

*Plant treatments*

We tested whether single terpene components attracted female egg parasitoids when offered in combination with odour from pine twigs that were not induced by egg deposition, but artificially wounded. Artificially wounded pine twigs were taken as background odour because the chemical analyses had compared the headspace of oviposition-induced twigs and artificially wounded ones (Mumm et al. 2003). Oviposition-induced twigs were not compared just with untreated twigs since wounding of pine needles which occurs also naturally during egg deposition might change the composition of volatiles. However, the behavioural bioassays had revealed that wounded twigs did not attract the egg parasitoids (Hilker et al. 2002a). Thus, a comparison of the headspace of wounded twigs and oviposition-induced ones made it possible to exclude all components released due to the pure wounding as potential key components causing the attractiveness.

For our experiments, small pine twigs (10-12 cm) were cut and provided with tap water. The treatment of the artificially wounded pine twigs were conducted as described by Mumm et al. (2003). Eight pine needles of a twig were slit tangentially with a clean insect needle. Wounded pine twigs were placed in water for 72 h at 25 °C, 18L:6D and 60% relative humidity. Prior to the bioassay, a ca. 5 cm long part of the twigs was cut and tightly wrapped with Parafilm®. This part of the twig was placed into the odour source flask of the olfactometer.

*Response of egg parasitoids to (E)-β-farnesene*

(*E*)-β-Farnesene were diluted in n-hexane (Roth, Karlsruhe, Germany). In all experiments, 0.1 ml of the diluted (*E*)-β-farnesene was applied to a piece of filter paper (approx. 16 cm$^2$). After 90 s, when most of the solvent had evaporated, the filter paper was placed into a glass flask (250 ml) of the olfactometer. Purified air flowing into the test field of the olfactometer passed through the flask which contained the odour sample. A filter paper with 0.1 ml hexane was placed into glass jars providing the odour for each of the three control fields (1-3). A female egg parasitoid was then introduced into the olfactometer. New filter papers with terpenes or hexane were used for each parasitoid. Three different concentrations were tested: 0.01, 0.1 and 1μg/μl hexane.

*Response of egg parasitoids to single terpenes in combination with pine volatiles*

In this experiment, we assessed the responses of egg parasitoids to (E)-β-farnesene and two other sesquiterpenes (β-caryophyllene, δ-cadinene) that were offered in combination with pine volatiles. Pine twigs were artificially wounded as described above. Volatiles of artificially wounded pine twigs were mixed with single sesquiterpenes and offered in the test field of the olfactometer. For this setup, the olfactometer was modified as follows: The air flow of the test field was split and was passed through a glass flask (250 ml) containing the artificially wounded pine twig and through a second flask which contained the filter paper with a terpene. After the air had passed the two sample flasks, the air flow was joined together again. Thus, volatiles of the pine twig and the applied terpenes were mixed just before entering the arena of the olfactometer. Control fields were provided with 0.1 ml hexane. A volume of 0.1 ml of the tested terpenes or solvent was applied to filter paper disks. (E)-β-Farnesene was tested in concentrations of 0.01, 0.1 and 1μg/μl hexane, (E)-β-caryophyllene and δ-cadinene were tested in concentrations of 0.1 and 1μg/μl hexane. New filter paper disks with terpenes or hexane were used for each parasitoid. Each artificially wounded pine twig was changed after 8-10 parasitoids tested.

*Data analysis*

Data were statistically evaluated using the Friedman ANOVA and the Wilcoxon-Wilcox test for multiple comparisons (Köhler et al. 1995) using the software program SPSS 11.0. (SPSS Inc., USA).

**Results**

*Response of egg parasitoids to (E)-β-farnesene*

Female egg parasitoids showed no significant response to none of the applied concentrations (0.01, 0.1 and 1µg/µl) of (E)-β-farnesene (Tab. 1). Therefore, (E)-β-farnesene *per se* was not attractive to the egg parasitoids at the concentrations tested.

*Response of egg parasitoids (E)-β-farnesene in combination with pine volatiles*

In order to elucidate whether *C. ruforum* responds to (E)-β-farnesene together with pine volatiles as background odour, we offered (E)-β-farnesene in combination with volatiles of artificially wounded pines. Volatiles of artificially wounded pine twigs were *per se* not attractive for female *C. ruforum* (Hilker et al. 2002a). When offered together with volatiles from artificially wounded twigs, (E)-β-farnesene applied at a concentration of 0.01 µg/µl elicited no significant response in *C. ruforum*. In contrast, the tenfold higher concentration (0.1µg/µl) of (E)-β-farnesene combined with pine volatiles significantly attracted egg parasitoids (Tab. 1). However, when (E)-β-farnesene was offered at a concentration of 1µg/µl on the background of volatiles from a wounded pine twig, the attractiveness switched to a repellent effect. Parasitoids avoided the test-field in the olfactometer compared to the control fields (Tab. 1). Thus, there is a dose-dependent effect of (E)-β-farnesene combined with volatiles of artificially wounded pine.

*Response of egg parasitoids to single sesquiterpenes in combination with pine volatiles*

To test whether the parasitoids' response to (E)-β-farnesene in combination with pine volatiles is specific for this compound, we used two other sesquiterpenes present in the headspace of *P. sylvestris,* i.e. (E)-β-caryophyllene and δ-cadinene. The sesquiterpenes (E)-β-caryophyllene and δ-cadinene were tested in combination with volatiles from wounded pine twigs at the concentrations at which (E)-β-farnesene was active (0.1µg/µl and 1µg/µl). The egg parasitoid *C. ruforum* did not respond to the concentration of 0.1µl/µg of (E)-β-caryophyllene in the olfactometer (Tab. 2). δ-

Cadinene tended to attract *C. ruforum* at the concentration of 0.1µl/µg, but this was not statistically significant (Tab. 2). The concentration of 1µg/µl of (*E*)-β-caryophyllene or δ-cadinene had neither an attractive nor a repellent effect to the egg parasitoids (Tab. 2).

## Discussion

Females of the egg parasitoid *C. ruforum* were not attracted by different concentrations of (*E*)-β-farnesene alone. However, when this component was offered on the background of a non-attractive natural blend of pine volatiles, this combination became attractive when tested at the intermediate concentration. The combination became repellent, when tested with a high concentration of (*E*)-β-farnesene. When combining the non-attractive natural blend of pine volatiles with other terpenoid components than (*E*)-β-farnesene, no such effects were detected at neither concentration tested.

Rutledge (1996) gives an overview of studies which show that single volatile constituents of the host plant were able to attract parasitoids. (*E*)-β-Farnesene is a common sesquiterpene that is released by many plants and also herbivorous insects, e.g. aphids (Nault and Bowers 1974). Numerous studies have shown increased amounts or *de novo* production of this sesquiterpene in plants induced by feeding herbivores (e.g. Bolter et al. 1997; De Moraes et al. 1998; Gols et al. 1999; Paré et al. 1999; Röse and Tumlinson 2004; Turlings et al. 1998), by mechanical damage (e.g. McAuslane and Alborn 1998), and by treatment with JA (e.g. Rodriguez-Saona et al. 2001; Schmelz et al. 2001). Numerous insect species are able to respond to (*E*)-β-farnesene. Herbivorous insects have been shown to respond towards (*E*)-β-farnesene (e.g. Bengtsson et al. 2001; Koshier et al. 2000). Both predators and parasitoids show EAG responses (e.g. Al Abassi et al. 2000; Du et al. 1998; Weissbecker et al. 2000; Zhu et al. 1999) as well as behaviourally responses towards (*E*)-β-farnesene (Du et al. 1998; Francis et al. 2004; Micha and Wyss 1996). However, in our study we could not show any behavioural response of *C. ruforum* to (*E*)-β-farnesene *per se*.

**Table 1.** Response of female egg parasitoids *C. ruforum* to different concentrations of synthetic (*E*)-β-farnesene alone (left) and offered in combination with non-induced volatiles of *Pinus sylvestris* twigs (right). Median values and interquartile range (parentheses) of the time parasitoid females spent in test (T) and control fields (1–3) of a four-arm-olfactometer are given. Test field with 0.1 ml of (*E*)-β-farnesene or 0.1 ml of (*E*)-β-farnesene with pine volatiles; (*E*)-β-farnesene was applied on filter paper; 1, 2, 3, = three control fields with 0.1 ml of hexane (solvent) applied on filter paper. n.s. indicates a non-significant ($P>0.05$) difference evaluated by a Friedman ANOVA. Different letters indicate significant ($P<0.05$) differences evaluated by the Wilcoxon–Wilcox test.

**(*E*)-β-farnesene**

| | (*E*)-β-farnesene walking duration [s] | | | | | |
|---|---|---|---|---|---|---|
| | T | 1 | 2 | 3 | N | |
| 0.01μg/μl | 103 (75-214) | 125 (61-210) | 114 (27-214) | 48 (24-165) | 20 | n.s. |
| 0.1μg/μl | 58 (26-160) | 71 (33-227) | 98 (62-263) | 79 (58-156) | 20 | n.s. |
| 1μg/μl | 98 (47-157) | 73 (17-180) | 95 (48-148) | 147 (49-221) | 22 | n.s. |

**(*E*)-β-farnesene combined with pine volatiles**

| | walking duration [s] | | | | | |
|---|---|---|---|---|---|---|
| | T | 1 | 2 | 3 | N | |
| 0.01μg/μl | 100 (49-201) | 158 (74-261) | 101 (58-185) | 90 (21-141) | 27 | n.s. |
| 0.1μg/μl | 141[a] (94-282) | 112[ab] (36-210) | 37[b] (13-152) | 24[b] (0-76) | 21 | p=0.003 |
| 1μg/μl | 78[a] (15-133) | 113[ab] (51-226) | 155[b] (114-197) | 92[ab] (34-134) | 21 | p=0.007 |

Synergistic effects of $(E)$-$\beta$-farnesene with other infochemicals mediating insect – plant interactions are well-known. For example, $(E)$-$\beta$-farnesene emitted by apples (*Malus* spp.) acts as synergist by significantly enhancing the attractiveness of the sex pheromone of *Cydia pomonella* (codlemone) for male codling moths (Yang et al. 2004). On the other hand, the response of the aphid *Lipaphis erysimi* to its alarm-pheromone $(E)$-$\beta$-farnesene was significantly increased, when $(E)$-$\beta$-farnesene was combined with plant-derived isocyanates (Dawson et al. 1987). However, in all these cases at least two different compounds that originate from different trophic levels (host plant and herbivore) were combined and then act in an additive or synergistic way.

**Table 2.** Response of female egg parasitoids *C. ruforum* to different concentrations of synthetic sesquiterpenes $(E)$-$\beta$-caryophyllene and $\delta$-cadinene offered in combination with non-induced volatiles of *Pinus sylvestris* twigs. Median values and interquartile range (parentheses) of the time parasitoid females spent in test (T) and control fields (1–3) of a four-arm-olfactometer are given. Test field with 0.1 ml of the sesquiterpenes with pine volatiles; $(E)$-$\beta$-farnesene was applied on filter paper; 1, 2, 3, = three control fields with 0.1 ml of hexane (solvent) applied on filter paper. n.s. indicates a non-significant ($P>0.05$) difference evaluated by a Friedman ANOVA.

| | | sesquiterpenes combined with non-induced pine volatiles | | | | | |
|---|---|---|---|---|---|---|---|
| | | walking duration [s] | | | | | |
| **0.1µg/µl** | | **T** | **1** | **2** | **3** | **N** | **statistics** |
| | $(E)$-$\beta$-caryophyllene | 130 (73-216) | 72 (6-147) | 44 (13-122) | 103 (12-268) | 25 | n.s. (p=0.24) |
| | $\delta$-cadinene | 196 (77-243) | 101 (51-195) | 101 (41-170) | 140 (60-194) | 31 | n.s. (p=0.08) |
| **1µg/µl** | | | | | | | |
| | $(E)$-$\beta$-caryophyllene | 144 (59-229) | 122 (46-189) | 95 (13-145) | 101 (8-169) | 22 | n.s. (p=0.40) |
| | $\delta$-cadinene | 153 (66-292) | 148 (80-203) | 95 (54-111) | 98 (58-216) | 23 | n.s. (p=0.24) |

This is different from our experiments where qualitatively no novel volatile mixture was composed, but only the ratio of a single compound ($(E)$-$\beta$-farnesene) to the whole blend was changed. $(E)$-$\beta$-Farnesene is present in untreated pine twigs, in artificially damaged ones, in JA-treated ones, and in oviposition-induced twigs. Only the two latter mentioned types of pine twig samples emit odour that is attractive to the egg parasitoid (Mumm et al. 2003). Thus, the attractive effect of the combination of (*E*)-β-farnesene at its intermediate concentration and the natural volatile blend of a non-attractive pine twig is not due to a synergism *sensu stricto* because neither the pure (*E*)-β-farnesene nor the volatiles of artificially wounded pines were attractive *per se* to the parasitoids.

Recently, two further tritrophic studies showed the importance of plant background odour when testing the behavioural response of carnivorous arthropods to single components of the natural plant headspace or to a mixture of specific components of the natural blend: (1) A synthetic mixture of volatiles composed of constituents of the headspace of bark beetle infested Norway spruce logs (*Picea abies*) was only as attractive as infested logs for bark beetle parasitoids when this mixture was offered at the background of odour from uninfested logs. Neither the synthetic mixture nor the odour of uninfested logs were attractive *per se* (Pettersson 2001; Pettersson et al. 2001). (2). Lima bean plants (*Phaseolus lunatus*) infested by spider mites (*Tetranychus urticae*) are known to produce methylsalicylate, thus attracting the predatory mites (de Boer and Dicke 2004). Methylsalicylate at low and intermediate concentrations did not attract predatory mites. Nor did volatiles from uninfested lima bean plants. However, when methylsalicylate was offered at the background of odour from uninfested plants, this combination attracted the predatory mites.

The attractiveness of (*E*)-β-farnesene mixed with non-oviposition-induced pine volatiles was dose-dependent. Only the concentration of 100ng/µl (*E*)-β-farnesene was able to attract *C. ruforum* females, whereas the concentration of 10ng/µl was not. Electrophysiological studies of the antennal olfactory system of *Musca domestica* revealed that background odour could increase the response to low concentrated odour pulses, whereas responses to higher concentrated volatile stimuli decreased when background odour was present (Kelling et al. 2002). Up to now, we do not know the physiological mechanisms responsible for the behavioural effects that we have detected here. Our results show that background odour was essential for the egg parasitoid to respond behaviourally to (*E*)-β-farnesene. Our data suggest that

*C. ruforum* is comparing the ratio of specifically this sesquiterpene to the other pine volatiles. A special ratio of $(E)$-β-farnesene within a pine volatile background might "tell" the egg parasitoid where to find host eggs. If the ratio is "wrong", i.e. either too low as in the experiment with 10ng/μl or too high (1μg/μl $(E)$-β-farnesene), parasitoids may not show a response or as in the latter case were even repelled. Further studies need to investigate whether the whole pine volatile bouquet is necessary to elicit the behavioural response of the egg parasitoids or whether a combination of single (key) components with $(E)$-β-farnesene elicit adequate responses in the parasitoids as suggested for other parasitoids (Dicke and Vet 1999; Vet et al. 1998).

## Acknowledgements

Special thanks are to Frank Müller for his help with the synthesis of $(E)$-β-farnesene. This study was supported by the Deutsche Forschungsgemeinschaft (DFG Hi 416/11-1,2).

## References

Al Abassi, S, Birkett, MA, Pettersson, J, Pickett, JA, Wadhams, LJ, Woodcock, CM (2000). Response of the seven-spot ladybird to an aphid alarm pheromone and an alarm pheromone inhibitor mediated by paired olfactory cells. J. Chem. Ecol. 26:1765-1771.

Bengtsson, M, Backmann, A-C, Ansebo, L, Anderson, P, Lofqvist, J, Witzgall, P (2001). Plant odor analysis of apple: Antennal response of codling moth females to apple volatiles during phenological development. J. Agric. Food Chem. 49:3736-3741.

Bolter, CJ, Dicke, M, van Loon, J-J-A, Visser, J-H, Posthumus, MA (1997). Attraction of Colorado potato beetle to herbivore-damaged plants during herbivory and after its termination. J. Chem. Ecol. 23:1003-1023.

Bombosch, S, Ramakers, PMJ (1976). Zur Dauerzucht von *Gilpinia hercyniae* Htg. Z. Pflanzenkrank Pflanzen 83:40-44.

Dawson, GW, Griffiths, DC, Pickett, JA, Wadhams, LJ, Woodcock, CM (1987). Plant-derived synergists of alarm pheromone from turnip aphid, *Lipahis* (*Hyadaphis*) *erysimi* (Homoptera, Aphididae). J. Chem. Ecol. 13:1663-1671.

de Boer, JG, Dicke, M (2004). The role of methyl salicylate in prey searching behavior of the predatory mite *Phytoseiulus persimilis*. J. Chem. Ecol. 30:255-271.

de Bruxelles, GL, Roberts, MR (2001). Signals regulating multiple responses to wounding and herbivores. Crit. Rev. Plant. Sci. 20:-487.

De Moraes, CM, Lewis, WJ, Paré, PW, Alborn, HT, Tumlinson, JH (1998). Herbivore-infested plants selectively attract parasitoids. Nature 393:570-573.

Dicke, M (1999). Evolution of induced indirect defense of plants. In: Tollrian, R, Harvell, CD (eds), The Ecology and Evolution of Inducible Defenses. Princeton University Press, Princeton, pp. 62-88.

Dicke, M, Hilker, M (2003). Induced plant defences: from molecular biology to evolutionary ecology. Basic Appl. Ecol. 4:3-14.

Dicke, M, van Poecke, RMP (2002). Signalling in plant-insect interactions: signal transduction in direct and indirect plant defence. In: Scheel, D, Wasternack, C (eds), Plant Signal Transduction. Oxford University Press, Oxford, pp. 289-316.

Dicke, M, van Loon, JJA (2000). Multitrophic effects of herbivore-induced plant volatiles in an evolutionary context. Entomol. Exp. Appl. 37:237-249.

Dicke, M, Vet, LEM (1999). Plant-carnivore interactions: evolutionary and ecological consequences for plant, herbivore and carnivore. In: Olff, H, Brown, VK, Drent, RH (eds), Herbivores: Between Plants and Predators. Blackwell Science, pp. 483-520.

Du, Y, Poppy, GM, Powell, W, Pickett, JA, Wadhams, LJ, Woodcock, CM (1998). Identification of semiochemicals released during feeding that attract Parasitoid *Aphidius ervi*. J. Chem. Ecol. 24:1355-1368.

Eichhorn, O (1976). Dauerzucht von *Diprion pini* L. (Hym.: Diprionidae) im Laboratorium unter Berücksichtigung der Fotoperiode. Anz. Schädlingskd. Pfl 49:38-41.

Francis, F, Lognay, G, Haubruge, E (2004). Olfactory responses to aphid and host plant volatile releases: (*E*)-β-farnesene an effective kairomone for the predator *Adalia bipunctata*. J. Chem. Ecol. 30:741-755.

Gatehouse, JA (2002). Plant resistance towards insect herbivores: a dynamic interaction. New Phytol. 156:145-169.

Gols, R, Posthumus, MA, Dicke, M (1999). Jasmonic acid induces the production of gerbera volatiles that attract the biological control agent *Phytoseiulus persimilis*. Entomol. Exp. Appl. 93:77-86.

Hilker, M, Meiners, T (2002). Induction of plant responses towards oviposition and feeding of herbivorous arthropods: a comparison. Entomol. Exp. Appl. 104:181-192.

Hilker, M, Kobs, C, Varama, M, Schrank, K (2002a). Insect egg deposition induces *Pinus* to attract egg parasitoids. J. Exp. Biol. 205:455-461.

Hilker, M, Rohfritsch, O, Meiners, T (2002b). The plant's response towards insect egg deposition. In: Hilker, M, Meiners, T (eds), Chemoecology of Insect Eggs and Egg Deposition. Blackwell Publishing, Berlin, Oxford, pp. 205-233.

Kang, S-K, Chung, G-Y, Lee, D-H (1987). A convenient synthesis of (*E*)-β-farnesene. Bull. Korean Chem. Soc. 8:351-353.

Kelling, FJ, Ialenti, F, Den Otter, CJ (2002). Background odour induces adaptation and sensitization of olfactory receptors in the antennae of houseflies. Med. Vet. Entomol. 16:161-169.

Koshier, EH, de Kogel, WJ, Visser, JH (2000). Assessing the attractiveness of volatile plant compounds to western flower thrips *Frankliniella occidentalis*. J. Chem. Ecol. 26:2643-2655.

Köhler, W, Schachtel, G, Voleske, P (1995). Biostatistik. Springer, Berlin.

McAuslane, HJ, Alborn, HT (1998). Systemic induction of allelochemicals in glanded and glandless isogenic cotton by *Spodoptera exigua* feeding. J. Chem. Ecol. 24:399-416.

Micha, SG, Wyss, U (1996). Aphid alarm pheromone (*E*)-β-farnesene: a host finding kairomone for the aphid primary parasitoid *Aphidius uzbekistanicus* (Hymenoptera: Aphidiinae). Chemoecology 7:132-139.

Moore, GE, Clark, EW (1968). Suppressing microorganisms and maintaining turgidity in coniferous foliage used to rear insects in the laboratory. J. Econ. Entomol. 61:1030-1031.

Mumm, R, Schrank, K, Wegener, R, Schulz, S, Hilker, M (2003). Chemical analysis of volatiles emitted by *Pinus sylvestris* after induction by insect oviposition. J. Chem. Ecol. 29:1235-1252.

Nault, LR, Bowers, WS (1974). Multiple alarm pheromones in aphids. Entomol. Exp. Appl. 17:455-457.

Paré, PW, Tumlinson, JH (1999). Plant volatiles as a defense against insect herbivores. Plant Physiol. 121:325-331.

Pettersson, EM (2001). Volatile attractants for three pteromalid parasitoids attacking concealed spruce bark beetles. Chemoecology 11:89-95.

Pettersson, EM, Birgersson, G, Witzgall, P (2001). Synthetic attractants for the bark beetle parasitoid *Coeloides bostrichorum* Giraud (Hymenoptera: Braconidae). Naturwissenschaften 88:88-91.

Pettersson, J (1970). An aphid sex attractant. I. Biological studies. Entomol. Scand. 1:63-73.

Rodriguez-Saona, C, Crafts-Brandner, SJ, Paré, PW, Henneberry, TJ (2001). Exogenous methyl jasmonate induces volatile emissions in cotton plants. J. Chem. Ecol. 27:679-695.

Röse, U, Tumlinson, JH (2004). Volatiles released from cotton plants in response to *Helicoverpa zea* feeding damage on cotton flower buds. Planta 218:824-832.

Rutledge, CE (1996). A survey of identified kairomones and synomones used by insect parasitoids to locate and accept their hosts. Chemoecology 7:121-131.

Schaller, F, Weiler, EW (2002). Wound- and mechanical signalling. In: Scheel, D, Wasternack, C (eds), Plant Signal Transduction. Oxford University Press, Oxford, pp. 20-44.

Schmelz, EA, Alborn, HT, Tumlinson, JH (2001). The influence of intact-plant and excised-leaf bioassay designs on volicitin- and jasmonic acid-induced sesquiterpene volatile release in *Zea mays*. Planta 214:171-179.

Smith, BH (1998). Analysis of interaction in binary mixtures. Physiol. Behav. 65:397-407.

Turlings, TCJ, Gouinguené, S, Degen, T, Fritzsche-Hoballah, ME (2002). The chemical ecology of plant-caterpillar-parasitoid interactions. In: Tscharntke, T, Hawkins, BA (eds), Multitrophic Level Interactions. Cambridge University Press, Cambridge, pp. 148-173.

Turlings, TCJ, Bernasconi, M, Bertossa, R, Bigler, F, Caloz, G, Dorn, S (1998). The induction of volatile emissions in maize by three herbivore species with different feeding habitats: possible consequences for their natural enemies. Biol. Control 11:122-129.

Turlings, TCJ, Wäckers, FL, Vet, LEM, Lewis, WJ, Tumlinson, JH (1993). Learning of host-finding cues by hymenopterous parasitoids. In: Papaj, DR, Lewis, AC (eds), Insect Learning. Ecological and Evolutionary Perspectives. Chapman & Hall, New York, pp. 51-78.

Vet, LEM, De Jong, AG, Franchi, E, Papaj, DR (1998). The effect of complete versus incomplete information on odour discrimination in a parasitic wasp. Anim. Behav. 55:1271-1279.

Vet, LEM, Lewis, WJ, Cardé, RT (1995). Parasitoid foraging and learning. In: Cardé, RT, Bell, WJ (eds), Chemical Ecology of Insects 2. Chapman and Hall, London; New York, pp. 65-101.

Vet, LEM, Van Lenterern, JC, Heymans, M, Meelis, E (1983). An airflow olfactometer for measuring olfactory responses of hymenopterous parasitoids and other small insects. Physiol. Entomol. 8:97-106.

Weissbecker, B, van Loon, JJA, Posthumus, MA, Bouwmeester, HJ, Dicke, M (2000). Identification of volatile potato sesquiterpenoids and their olfactory detection by the two-spotted stinkbug *Perillus bioculatus*. J. Chem. Ecol. 26:1433-1445.

Yang, Z, Bengtsson, M, Witzgall, P (2004). Host plant volatiles synergize response to sex pheromone in codling moth, *Cydia pomonella*. J. Chem. Ecol. 30:619-629.

Zhu, J, Cossé, AA, Obrycki, JJ, Boo, KS, Baker, TC (1999). Olfactory reactions of the twelve-spotted lady beetle, *Coleomegilla maculata* and the green lacewing, *Chrysoperla carnea* to semiochemicals released from their prey and host plant: eletroantennogram and behavioral responses. J. Chem. Ecol. 25:1163-1177.

# Chapter 5

# Choosy Egg Parasitoids: Specificity of Oviposition-Induced Pine Volatiles Exploited by an Egg Parasitoid of Pine Sawflies

**Abstract** Generalist parasitoids are well-known to cope with the high genotypic and phenotypic plasticity of plant volatiles by learning odours during host encounter. In contrast, specialised parasitoids often respond innately to host specific cues. Previous studies have shown that females of the specialised egg parasitoid *Chrysonotomyia ruforum* (Hymenoptera, Eulophidae) are attracted to volatiles from *Pinus sylvestris* L. induced by egg deposition of its host *Diprion pini* L. (Hymenoptera, Diprionidae) when they previously experienced pine twigs with host eggs. In this study, we investigated by olfactometer studies how specifically *C. ruforum* is responding to oviposition-induced plant volatiles. Furthermore, we studied whether parasitoids show an innate response to oviposition-induced pine volatiles. Naïve parasitoids were not attracted to oviposition-induced pine volatiles. The attractiveness of volatiles from pines carrying eggs was shown to be specific for the pine and herbivore species, respectively (species specificity). We also tested, whether not only oviposition, but also larval feeding induces attractive volatiles (developmental stage specificity). Feeding of *D. pini* larvae did not induce the emission of *P. sylvestris* volatiles attractive to the egg parasitoid. Our results show that a specialist egg parasitoid does not innately show a positive response to oviposition-induced plant volatiles, but needs to learn them. Furthermore, the results show that *C. ruforum* as a specialist obviously is not able to learn a wide range of volatiles like some generalists are, but instead learns only a very specific oviposition-induced plant volatile pattern, i.e. a pattern induced by the most preferred host species laying eggs on the most preferred food plant.

**Keywords** foraging behaviour, learning, indirect plant defence, induced plant defence, infochemical

**Introduction**

Parasitoids of herbivores are known to use plant volatiles when foraging for hosts (e.g. Vinson 1991; Vet and Dicke 1992; Rutledge 1996; Steidle and van Loon 2003). Herbivore damage caused by feeding or egg deposition may induce qualitative and quantitative changes in the plant's volatile blend (e.g. Dicke 1994; Turlings and Fritzsche 1999; Dicke and van Loon 2000; Hilker and Meiners 2002). Such induced plant volatiles may be specific for a herbivore species (e.g. De Moraes et al. 1998; Du et al. 1998; Turlings et al. 1998, 2002) or a particular developmental stage of the herbivore (e.g. Takabayashi et al. 1995; Gouinguené et al. 2003). Moreover, even related plant species or conspecific varieties attacked by the same herbivore species emit specific volatile patterns (Takabayashi and Dicke 1996; Geervliet et al. 1997; Gouinguené et al. 2001). The ability of parasitoids to discriminate between specific plant volatile blends has been reported for numerous tritrophic systems (reviewed by Dicke 1999; Fritzsche-Hoballah et al. 2002).

The high geno- and phenotypic variability of herbivore-induced plant volatiles might limit the reliability of these cues for parasitoids to find a suitable host (reviewed by Dicke and Vet 1999; Dicke and Hilker 2003). Parasitoids show either a fixed innate response to specific cues or deal with this variability by learning odours associatively during a host encounter thus enabling them to adjust or reinforce their responses to changes in host-related cues (Turlings et al. 1993; Vet et al. 1995). Specialised parasitoids, attacking just few herbivore species that feed upon only few plant species, are generally supposed to respond to infochemicals innately and to use primarily specific cues for finding a host (Vet and Dicke 1992; Steidle and van Loon 2003).

In a previous study, Hilker et al. (2002) demonstrated that the specialised egg parasitoid *Chrysonotomyia ruforum* (Krausse) (Hymenoptera, Eulophidae) was attracted to volatiles from *Pinus sylvestris* L. (Pinales, Pinaceae) induced by egg deposition of the herbivorous sawfly *Diprion pini* L. (Hymenoptera, Diprionidae). The host range of *C. ruforum* is restricted to eggs of members of the subfamily

Diprioninae (Pschorn-Walcher and Eichhorn 1973; Eichhorn and Pschorn-Walcher 1976) that mainly feed on pines (*Pinus* spp.) in Europe (Pschorn-Walcher 1982).

In the study presented here, we investigated the specificity of chemical cues used by the egg parasitoid *C. ruforum* for host search. Firstly, we investigated whether female egg parasitoids respond innately to oviposition-induced pine volatiles. Secondly, we tested whether the induction of plant volatiles by egg deposition is specific for the plant or herbivore species (species specificity). In addition, we studied whether larval feeding of *D. pini* also induces volatiles in *P. sylvestris* that attract female *C. ruforum,* as eggs of *D. pini* do (developmental stage specificity).

To examine the plant specificity of this tritrophic interaction, we tested whether volatiles from the Austrian black pine (*Pinus nigra* Arnold var. *nigra*) carrying eggs of *D. pini* attract *C. ruforum* females. It has been reported that *D. pini* accepts *P. nigra* as a host for oviposition (Eliescu 1932; Zivojinovic 1954; Auger et al. 1994; Barre et al. 2002). *P. nigra* var. *nigra* is naturally distributed mainly in southeastern Europe (Austria, Italy, Balkan Peninsula) but has been cultivated for parks and forests worldwide (Krüssmann 1983; Schütt et al. 1992; Rafii et al. 1996).

Herbivore specificity was studied by testing the response of *C. ruforum* to volatiles from *P. sylvestris* twigs on which eggs of *Gilpinia pallida* Klug or *Neodiprion sertifer* Geoffroy were deposited. The diprionid species are frequently occurring herbivorous sawflies of Scots pine (*P. sylvestris*) (Pschorn-Walcher 1982). Eggs of *N. sertifer* and *G. pallida* are both known as suitable hosts for *C. ruforum* (Pschorn-Walcher and Eichhorn 1973; Pschorn-Walcher 1988).

**Materials and methods**

*Plants and insects*

Branches of *Pinus sylvestris* used for experiments and insect rearing were detached from crowns of 15- to 35-year-old trees in the forests near Berlin. Branches from *Pinus nigra* var. *nigra* were cut from a 14-year-old pine stand near Berlin. For the experiments, small twigs were detached from these branches (see below for further details). For the rearing of sawflies, all stems were cleaned and sterilized according to the method of Moore and Clark (1968).

*Diprion pini* was reared continuously in the laboratory on cut pine twigs, as described by Bombosch and Ramakers (1976) and Eichhorn (1976) at $25\pm1$ °C, L/D 18:6 h, and 70 % relative humidity. The laboratory culture of *D. pini* was started with cocoons collected in the field in France (by C. Géri, INRA, Orléans). Adults of *Gilpinia pallida* were reared from cocoons (collected in the field in Finland) under the same conditions as described for *D. pini*. *Neodiprion sertifer* adults emerged from cocoons collected by A. Martini (Univ. of Bologna, Italy) from pine trees in Italy. A continuous laboratory rearing of this species was not established.

The egg parasitoid *Chrysonotomyia ruforum* was obtained from parasitized eggs of *D. pini* and *N. sertifer* collected in the field in France (near Fontainebleau) and southern and central Finland. Parasitized eggs were kept in Petri dishes (i.d. 9 cm) in a climate chamber at 10 °C, 18L:6D photoperiod, and 70 % relative humidity. To induce parasitoid emergence, needles with parasitized eggs were placed in a climate chamber at 25 °C, 18L:6D photoperiod and 70 % relative humidity. Emerging adults were collected daily and transferred in small perspex tubes (75mm long, 15mm i.d.) covered with gauze at one end. A cotton-wool plug moistened with an aqueous honey solution closed the other end. About 20 male and 20 female parasitoids were confined per tube. Mating was normally observed soon after emergence. The parasitoids were kept at 10 °C, 18L:6D until they were used for bioassays. Parasitoids used for bioassays were 2-16 days old.

To test whether the response to volatiles from differently treated pine twigs was dependent on the plant-host complex experienced by the parasitoids, two types of treatments were investigated: (*i*) Parasitoids were given the chance to experience the same plant-host complex that was offered in the olfactometer (within experiment), (*ii*) experienced and tested plant-host complex were different (cross experiment). A detailed scheme of the experiments is given in table 2. When testing experienced female parasitoids, these had prior contact with sawfly eggs on pine twigs for a period of 24 h. After this exposure time, female parasitoids were kept isolated from host eggs on pine twigs for 24 h in a Petri dish (compare Hilker et al. 2002). Naïve wasps had no contact to sawfly eggs prior to the experiments.

*Olfactometer bioassay – general procedures and data collection*

All bioassays were conducted in a four-arm olfactometer (Pettersson 1970; Vet et al. 1983) as described by Hilker et al. (2002) in detail. The airflow was adjusted to 155ml min$^{-1}$. When starting a bioassay, a parasitoid female was introduced into the arena of the olfactometer. We recorded how much time the parasitoid was present within each of the four odour fields over a period of 600 s using the Observer program 3.0 (Wageningen, The Netherlands). Data obtained from parasitoids that walked less than 300 s were discarded. For each treatment, 22-40 parasitoids and 4-6 plant samples were tested. Data were statistically evaluated using the Friedman ANOVA and the Wilcoxon-Wilcox test for multiple comparisons (Köhler et al. 1995) using the software program SPSS 11.0. (SPSS Inc., USA). We term an odour "attractive" when the parasitoid prefers walking in the olfactometer field provided with this odour since significantly longer walking periods in the odour field is usually interpreted as a response of the parasitoid to an attractive odour (Hilker et al. 2002).

*Plant treatments general*

Small pine twigs (10-15 cm) of *P. sylvestris* or *P. nigra* laden with host eggs were obtained similar to the method described by Hilker et al. (2002). Two female and two male sawflies were confined with the twigs. Females were allowed to mate and to lay eggs. After a period of 72 h, twigs carrying eggs were removed from the tap water in which they were kept. They were tightly wrapped with parafilm at the cut end when used for the bioassays.

*Species specificity*

In order to investigate whether the emission of attractive plant volatiles induced by oviposition is specific for the tritrophic system *P. sylvestris* - *D. pini* –

*C. ruforum*, we tested the attractiveness of volatiles from pine twigs subjected to the following treatments, changing either the plant or herbivore species or both:

*(a) Plant specificity.* To study how *C. ruforum* reacts to volatiles from egg-laden twigs other than from *P. sylvestris* (Hilker et al. 2002), the response of experienced parasitoid females (see above) to volatiles from twigs of *P. nigra* var. *nigra* carrying eggs of *D. pini* was tested. Twigs were treated as described above .

*(b) Herbivore specificity.* We tested whether egg deposition by *Gilpinia pallida* or *N. sertifer* induces *P. sylvestris* to emit volatiles that attract the egg parasitoid *C. ruforum*. Pine twigs which were carrying eggs of these species for a period of 72 h were used for the bioassays as described above.

*Developmental stage specificity*

To investigate whether not only oviposition, but also feeding of sawfly larvae induces emission of volatiles attractive to the egg parasitoids, two different bioassays were performed. First, 25 young *D. pini* larvae (L 1/2) were allowed to feed on small *P. sylvestris* twigs (ca. 10 cm long) for 24 h. Pine twigs were used for olfactometer bioassays immediately after removal of larvae. In contrast to the treatment period of 72 h used for oviposition-exposed plants, we reduced the period of treatment in these feeding-exposed plants because the twigs would otherwise have been completely consumed.

In order to carry out an experiment after a treatment period of 72 h, we modified the set up as follows. We used the fact that oviposition-induced volatiles are not only emitted locally at the site of egg deposition, but also systemically in adjacent, egg-free parts of a twig after 72 h (Hilker et al. 2002). If larval feeding induces attractive volatiles in pine twigs, we hypothesized that this response would also be systemic. Therefore, we modified the treatment procedure and used a method for systemic induction described by Hilker et al. (2002): 25 young larvae of *D. pini* (L 1/2) were placed on the lower half of a pine twig (approx. 15 cm long), while the upper half of the twig was covered with polyethylene terephtalate (PET) foil to prevent feeding. The bag was ventilated with purified air through an in- and outlet. After a feeding period of 72 h, the upper half of the twig was cut and the foil was removed. The cut end of the upper twig was tightly wrapped with Parafilm®. The response of *C. ruforum* females to volatiles from the upper part of the twig was tested in the olfactometer.


**Results**

*Significance of parasitoid experience*

Naïve *C. ruforum* females were not attracted to *P. sylvestris* volatiles induced by oviposition of *D. pini* (Tab. 1 A1). On the other hand, egg parasitoids that had previous experience with this plant-host complex were significantly attracted to oviposition-induced pine volatiles (Tab. 1 A2). Thus, a previous access to a plant-host complex was essential for *C. ruforum* to learn to respond to oviposition-induced pine volatiles. Therefore, all following experiments were conducted with experienced parasitoids only.

*Species specificity*

Volatiles from *P. nigra* with eggs of *D. pini* did not attract female *C. ruforum*, even though the parasitoids had experienced the same plant-host complex prior to the bioassay (Tab. 2 B1). Neither were volatiles from *P. sylvestris* carrying eggs of *G. pallida* attractive to female *C. ruforum* (Tab. 2 B2). In contrast, parasitoids were significantly attracted to volatiles from *P. sylvestris* induced by egg deposition of *N. sertifer* (Tab. 2 B3).


In cross experiments, the parasitoid had experienced a plant-host complex different from the one tested in the olfactometer assay. Parasitoids that had experienced odours from the *P. sylvestris - D. pini* complex did not respond significantly to volatiles from *P. nigra* carrying *D. pini* eggs (Tab. 2 C1). Nor did they show a significant response when they had experienced volatiles from the *P. nigra – D. pini* complex and were tested for their response to *P. sylvestris* twigs with eggs of *D. pini* (Tab. 2 C2). Parasitoids that had been exposed to volatiles from *P.*

*sylvestris* laden with eggs of the major host *D. pini* did not respond significantly to volatiles from *P. sylvestris* with eggs of *G. pallida* (Tab. 2 C3 and compare with B2). Volatiles from *P. sylvestris* with *D. pini* eggs were no longer attractive to the parasitoids, if they did not experience the major host prior to the bioassay (Tab. 2 C4).

Thus, *C. ruforum* only responded to *P. sylvestris* volatiles induced by egg deposition of the major hosts (*D. pini* and *N. sertifer*) when these hosts had been experienced prior to the bioassay.

*Developmental stage specificity*

Larval feeding for a period of 24 h did not locally induce the emission of volatiles in *P. sylvestris* twigs that attract the egg parasitoid *C. ruforum* (Friedman-ANOVA, $\chi^2$=3.57, p=0.311, N=23). Nor were parasitoids attracted to volatiles from undamaged twig parts when 25 larvae had fed on adjacent parts for 72 h (Friedman-ANOVA, p=0.896, N=22).

**Discussion**

The egg parasitoid *C. ruforum* did not show an innate response to volatiles from *P. sylvestris* induced by oviposition of *D. pini*. However, they were able to significantly response to those volatiles that had been experienced before and could be associated with the presence of host eggs (Tab 1 A1-2).

According to a literature survey by Steidle and van Loon (2003), learning behaviour of specialised parasitoids like *C. ruforum* has been shown in significantly fewer species compared to parasitoids with a broader foraging range. Instead, specialised parasitoids searching for a host were primarily shown to respond innately to specific chemical cues from the host, the host's plant, or from both lower trophic levels. The egg parasitoid *C. ruforum* is known to respond innately to the sex pheromones of its sawfly hosts (Hilker et al. 2000).

**Table 1** Parasitoid experience. Responses of naïve (A1) or experienced (A2) *Chrysonotomyia ruforum* females to volatiles from *Pinus* twigs induced by egg deposition of *D. pini* offered in a test field (Test) of a four-arm olfactometer. Control fields were supplied with clean air (1-3). The time the parasitoid females were present in the test and control fields are given over an observation period of 600s. Median values and interquartile ranges (parentheses) are given. *** indicates a significant (P<0.001) and n.s. a non-significant (P>0.05) difference evaluated by a Friedman ANOVA. Different letters indicate significant (P<0.001) differences evaluated by the Wilcoxon–Wilcox test.

| | | plant / host complex during experience | plant / host complex for bioassay | duration of stay [s] | | | | N | statistics |
|---|---|---|---|---|---|---|---|---|---|
| | | | | Test | 1 | 2 | 3 | | |
| **Naïve** | A1 | -- | *P. sylvestris –*<br>*D. pini* | 95<br>(24-283) | 113<br>(19-280) | 108<br>(33-326) | 86<br>(6-154) | 22 | n.s. (p=0.274) |
| **Experienced** | A2 | *P. sylvestris –*<br>*D. pini* | *P. sylvestris –*<br>*D. pini* | 387[a]<br>(331-437) | 45[b]<br>(15-98) | 33[b]<br>(10-62) | 116[b]<br>(52-166) | 27 | *** (p<0.001) |

Although these sex pheromones are highly reliable indicating habitats containing adult hosts, the detectability is low regarding the amount and especially the appearance in time (Vinson 1998; Steidle and van Loon 2002 for discussion). On the other hand, pine volatiles that are induced by egg deposition indicate the presence of host eggs, but may not be as reliable due to high qualitative and quantitative variation (Vet and Dicke 1992).

We suggest that learning of abundant oviposition-induced plant cues is beneficial for the specialised egg parasitoid *C. ruforum* and argue as follows:

*(a)*   Pines possess a high qualitative and quantitative variability in the composition of secondary terpenoid compounds between genotypes (Sjödin et al. 2000; Petrakis et al. 2001; Krauze-Baranowska et al. 2002), within trees (e.g. Barnola et al. 1997; Forrest et al. 2000; Latta et al. 2000) and individual tissues (e.g. Sjödin et al. 1996, 2000; Litvak and Monson 1998; Kleinhentz et al. 1999; Latta et al. 2000; Manninen et al. 2002). Furthermore, emission rates of terpenes are strongly affected by environmental factors such as temperature and light intensity (e.g. Tingey et al. 1991). Thus, an egg parasitoid searching for host eggs within a pine tree needs to cope with volatiles bouquets that might vary with the genotype, the position within the tree, the herbivores attacking the tree, the season, the daytime, and other environmental factors.

*(b)*   Mumm et al. (2003) could demonstrate that oviposition by *D. pini* on *P. sylvestris* does not result in qualitative but small quantitative changes in the pine volatile pattern. These results show that the egg parasitoid *C. ruforum* is able to detect very small changes of the pine volatile pattern after having experienced that this volatile pattern is associated with eggs of the major host. A fixed innate response to such small quantitative changes of the pine's volatile pattern induced by egg deposition might be not adaptive when taking into account that this pattern might vary in dependence of numerous other factors. Instead, learning could be a favourable strategy enabling *C. ruforum* to respond very flexibly to varying conditions and to adjust its response to finely tuned small quantitative volatile changes induced by host egg deposition.

The response of *C. ruforum* to oviposition-induced pine volatiles was shown to be specific for the plant species. Volatiles emitted from *P. sylvestris* after egg deposition of *D. pini* were attractive to *C. ruforum* (Tab. 1 A2), as was also shown in a previous study (Hilker et al. 2002). However, volatiles from *P. sylvestris* induced by egg deposition of *D. pini* were only attractive for *C. ruforum* if this plant-host complex had previously been experienced (Tab. 1 A2 and Tab. 2 C2). Surprisingly, *C. ruforum* females that had experience with the *P. nigra* - *D. pini* complex did not respond significantly to volatiles from the same plant-host complex (Tab. 2 B1). These results suggest that *C. ruforum* has not developed the ability to learn odours from *P. nigra* carrying eggs of *D. pini*.

The positive response of egg parasitoids to oviposition-induced *P. sylvestris* volatiles and their "non-response" to egg-carrying *P. nigra* twigs might be due to quantitative and qualitative differences between the volatile patterns of the two pine species (see Chapter 5). The headspace of egg-laden *P nigra* twigs might lack components that are necessary for *C. ruforum* to become attracted. Furthermore, the egg-carrying *P. nigra* twigs might release such a quantitative composition of volatiles that cannot be associated by *C. ruforum* with the presence of suitable hosts. Moreover, compounds emitted by egg-laden *P. nigra* may mask the attractiveness of other key compounds in the volatile mixture (Chandra and Smith 1998, Laloi et al. 2000, Meiners et al. 2003, and references therein).

*Pinus nigra* is known as a host plant of poor quality for *D. pini*. Egg development and larval performance of the sawfly is significantly reduced when compared to *P. sylvestris.* (Auger at al. 1994; Barre et al. 2002). Sawfly eggs laid on a suboptimal plant might be no preferred hosts for *C. ruforum* to forage for, since also the parasitoid's development might be negatively affected on such a plant , as was shown for other parasitoids (Turlings and Benrey 1998; Keasar et al. 2001; Sznajder and Harvey 2003; Takasu and Lewis 2003).

**Table 2** Species specificity of oviposition-induced plant volatiles. Responses of experienced (B1-3, C1-5) *Chrysonotomyia ruforum* females to volatiles from twigs of two *Pinus* species carrying eggs of different sawfly species offered in a test field (Test) of a four-arm olfactometer. Control fields were supplied with clean air (1-3). The time the parasitoid females were present in the test and control fields are given over an observation period of 600s. Median values and interquartile ranges (parentheses) are given. ** indicates a significant (P<0.01) and n.s. a non-significant (P>0.05) difference evaluated by a Friedman ANOVA. Different letters indicate significant (P<0.05) differences evaluated by the Wilcoxon–Wilcox test.

| | | plant / host complex during experience | plant / host complex for bioassay | duration of stay [s] | | | | N | statistics |
|---|---|---|---|---|---|---|---|---|---|
| | | | | Test | 1 | 2 | 3 | | |
| **Within experiments** | B1 | P. nigra – D. pini | P. nigra – D. pini | 95 (49-134) | 211 (87-295) | 186 (74-251) | 69 (20-177) | 25 | n.s. (p=0.069) |
| | B2 | P. sylvestris – G. pallida | P. sylvestris – G. pallida | 161 (73-216) | 107 (59-170) | 148 (54-242) | 139 (76-236) | 30 | n.s. (p=0.378) |
| | B3 | P. sylvestris – N. sertifer | P. sylvestris – N. sertifer | 209$^a$ (140-403) | 69$^{ab}$ (14-167) | 104$^b$ (0-176) | 83$^b$ (38-155) | 23 | ** (p=0.006) |
| **Cross experiments** | C1 | P. sylvestris – D. pini | P. nigra – D. pini | 92 (39-188) | 164 (39-188) | 136 (42-227) | 102 (29-186) | 27 | n.s. (p=0.39) |
| | C2 | P. nigra – D. pini | P. sylvestris – D. pini | 136 (68-206) | 147 (82-182) | 139 (82-227) | 116 (93-165) | 26 | n.s. (p=0.93) |
| | C3 | P. sylvestris – D. pini | P. sylvestris – G. pallida | 113 (82-196) | 160 (92-214) | 138 (87-220) | 96 (56-117) | 25 | n.s. (p=0.336) |
| | C4 | P. sylvestris – G. pallida | P. sylvestris – D. pini | 115 (55-232) | 101 (40-205) | 89 (44-189) | 176 (107-240) | 28 | n.s. (p=0.296) |

Also Meiners et al. (2000) showed that the response of the eulophid egg parasitoid *Oomyzus gallerucae* to elm volatiles induced by oviposition of its host, the elm leaf beetle, is specific for the plant species most favourable for the beetle.

The response of *C. ruforum* to oviposition-induced pine volatiles was specific for the herbivore species that laid eggs on *P. sylvestris*. Oviposition by *D. pini* and *N. sertifer* induced the emission of pine volatiles attractive to *C. ruforum*, whereas egg deposition by *G. pallida* did not. Egg depositions of closely related sawfly species obviously elicit different responses in *P. sylvestris*, which can be learnt by *C. ruforum* or not. The sawflies *D. pini* and *N. sertifer* represent major defoliators of *P. sylvestris* in Europe by causing severe damage to coniferous forests during an outbreak (Pschorn-Walcher 1982). On the other hand, *G. pallida* is considered to be only a marginal pest (Pschorn-Walcher 1982). Oviposition-induced responses might have evolved in *P. sylvestris* specific to major pest insects as a counteradaptation to the high damage caused by the insects. *C. ruforum* might have developed a specific ability to learn pine volatiles induced by oviposition of the favoured and most abundant host species in order to search for hosts at those sites where maximum host availability, and thus maximum reproduction, is possible.

To be able to respond in a differentiated manner, plants must be able to "recognise" the egg-laying species. Herbivore-borne elicitors are assumed to be responsible for the recognition process by plants (Stout and Bostock 1999). The elicitor inducing the production of attractive volatiles in *P. sylvestris* after egg deposition by *D. pini* is located in the oviduct secretion of the sawfly females (Hilker et al. 2002). Eggs are coated by the oviduct secretion when laid into pine needles. Further studies are needed to elucidate whether *P. sylvestris* "recognises" the ovipositing diprionid species by chemical differences between the oviduct secretions or by other means such as differences in wounding of the pine needles prior to egg deposition.

Egg parasitoids were not attracted to volatiles from pine twigs which were damaged by feeding sawfly larvae. Therefore, the response of the egg parasitoid *C.*

*ruforum* is specific for pine odours induced by a suitable developmental stage that can be parasitized. From the plant's perspective, the production of oviposition-induced volatiles seems to be a specific response to egg deposition and not a general reaction to herbivore damage. This specific plant response might be mediated by the elicitor in the oviduct secretion (see above). Furthermore, foraging *C. ruforum* that would rely on plant odours induced by larval feeding might be especially fooled because *D. pini* females avoided laying eggs on pine twigs treated with oral secretions of conspecific larvae and thus, sawfly females might move from larval infested sites to uninfested ones to avoid intraspecific competition (Hilker and Weitzel 1991).

In conclusion, the egg parasitoid *C. ruforum* specialized on diprionid hosts was shown to be able to learn cues specific for the plant species that is most beneficial for herbivore performance, for the herbivore species most abundant, and for the developmental stage (i.e. the egg stage) suitable for parasitization. Our results show that a specialist egg parasitoid does not innately respond to oviposition-induced plant volatiles, but is able to learn using those volatile patterns for host location that seem most beneficial.

**Acknowledgements**

We are very grateful to Dr. Antonio Martini (University of Bologna, Italy) for collecting *N. sertifer* in the field. Many thanks are due to Ute Braun who helped to rear *D. pini*. This study was supported by the Deutsche Forschungsgemeinschaft (DFG Hi 416/11-1,2).

# References

Auger, M-A, Géri, C, Allais, J-P (1994). Effect of the foliage of different pine species on the development and on the oviposition of the pine sawfly *Diprion pini* L. (Hym., Diprionidae); II. Influence on egg laying and interspecific variability of some active secondary compounds. J. Appl. Ent. 117:165-181.

Barnola, LF, Cedeno, A, Hasegawa, M (1997). Intraindividual variations of volatile terpene contents in *Pinus caribaea* needles and its possible relationship to *Atta laevigata* herbivory. Biochem. Syst. Ecol. 25:707-716.

Barre, F, Milsant, F, Palasse, C, Prigent, V, Goussard, F, Géri, C (2002). Preference and performance of the sawfly *Diprion pini* on host and non-host plants of the genus *Pinus*. Entomol. Exp. Appl. 102:229-237.

Bombosch, S, Ramakers, PMJ (1976). Zur Dauerzucht von *Gilpinia hercyniae* Htg. Z. Pflanzenkrank. Pflanzen. 83:40-44.

Chandra, S, Smith, BH (1998). An analysis of synthetic processing of odor mixtures in the honeybee (*Apis mellifera*). J. Exp. Biol. 201:3113-3121.

De Moraes, CM, Lewis, WJ, Paré, PW, Alborn, HT, Tumlinson, JH (1998). Herbivore-infested plants selectively attract parasitoids. Nature 393:570-573.

Dicke, M (1999). Are herbivore-induced plant volatiles reliable indicators of herbivore identity to foraging carnivorous arthropods? Entomol. Exp. Appl. 91:131-142.

Dicke, M (1994). Local and systemic production of volatile herbivore-induced terpenoids: their role in plant-carnivore mutualism. J. Plant Physiol. 143:465-472.

Dicke, M, Hilker, M (2003). Induced plant defences: from molecular biology to evolutionary ecology. Basic Appl. Ecol. 4 :3-14.

Dicke, M, van Loon, JJA (2000). Multitrophic effects of herbivore-induced plant volatiles in an evolutionary context. Entomol. Exp. Appl. 37:237-249.

Dicke, M, Vet, LEM (1999). Plant-carnivore interactions: evolutionary and ecological consequences for plant, herbivore and carnivore. In: Olff, H, Brown, VK, Drent, RH (eds), Herbivores: Between Plants and Predators. Blackwell Science, Oxford, pp. 483-520.

Du, Y, Poppy, GM, Powell, W, Pickett, JA, Wadhams, LJ, Woodcock, CM (1998). Identification of semiochemicals released during feeding that attract Parasitoid *Aphidius ervi*. J. Chem. Ecol. 24:1355-1368.

Eichhorn, O (1976). Dauerzucht von *Diprion pini* L. (Hym.: Diprionidae) im Laboratorium unter Berücksichtigung der Fotoperiode. Anz. Schädlingskd. Pfl. 49:38-41.

Eichhorn, O, Pschorn-Walcher, H (1976). Studies on the biology and ecology of the egg-parasites (Hym.: Chalcidoidea) of the pine sawfly *Diprion pini* (L.) (Hym.: Diprionidae) in Central Europe. Z. ang. Ent. 80:355-381.

Eliescu, G (1932). Beiträge zur Kenntnis der Morphologie, Anatomie und Biologie von *Lophyrus pini*. Z. ang. Ent. 19:22 (188)-67 (206).

Forrest, I, Burg, K, Klumpp, R (2000). Genetic markers: tools for identifying and characterising Scots pine populations. Invest. Agr. : Sist. Recur. For. : Fuera de Serie n. 1:67-88.

Fritzsche-Hoballah, ME, Tamò, C , Turlings, TCJ (2002). Differential attractiveness of induced odors emitted by eight maize varieties for the parasitoid *Cotesia marginiventris*: Is quality or quantity important? J. Chem. Ecol. 28:951-968.

Geervliet, JBF, Posthumus, MA, Vet, LEM, Dicke, M (1997). Comparative analysis of headspace volatiles from different caterpillar-infested or uninfested food plants of *Pieris* species. J. Chem. Ecol. 23:2935-2954.

Gouinguené, S, Alborn, H, Turlings, TCJ (2003). Induction of volatile emissions in maize by different larval instars of *Spodoptera littoralis*. J. Chem. Ecol. 29:145-162.

Gouinguené, S, Degen, T, Turlings, TCJ (2001). Variability in herbivore-induced odour emissions among maize cultivars and their wild ancestors (teosinte). Chemoecology 11:9-16.

Hilker, M, Meiners, T (2002). Induction of plant responses towards oviposition and feeding of herbivorous arthropods: a comparison. Entomol. Exp. Appl. 104:181-192.

Hilker, M, Weitzel, C (1991). Oviposition deterrence by chemical signals of conspecific larvae in *Diprion pini* (Hymenoptera: Diprionidae) and *Phyllodecta vulgatissima* (Coleoptera: Chrysomelidae). Entomol. Gen. 15:293-301.

Hilker, M, Kobs, C, Varama, M, Schrank, K (2002). Insect egg deposition induces *Pinus* to attract egg parasitoids. J. Exp. Biol. 205:455-461.

Hilker, M, Bläske, V, Kobs, C, Dippel, C (2000). Kairomonal effects of sawfly sex pheromones on egg parasitoids. J. Chem. Ecol. 26:2591-2601.

Keasar, T, Ney-Nifle, M, Mangel, M, Swenzey, S (2001). Early oviposition experience affects patch residence time in a foraging parasitoid. Entomol. Exp. Appl. 98:123-132.

Kleinhentz, M, Jactel, H, Menassieu, P (1999). Terpene attractant candidates of *Dioryctria sylvestrella* in maritime pine (*Pinus pinaster*) oleoresin, needles, liber, and headspace samples. J. Chem. Ecol. 25:2741-2756.

Köhler, W, Schachtel, G, Voleske, P (1995). Biostatistik. Springer, Berlin.

Krauze-Baranowska, M, Mardarowicz, M, Wiwart, M, Poblocka, L, Dynowska, M (2002). Antifungal activity of the essential oils from some species of the genus *Pinus*. Z. Naturforschung 57c:478-482.

Krüssmann, G (1983). Handbuch der Nadelgehölze. Parey, Berlin.

Laloi, D, Bailez, O, Blight, MM, Roger, B, Pham-Delegue, MH (2000). Recognition of complex odours by restrained and free-flying honey bees, *Apis mellifera*. J. Chem. Ecol. 26:2307-2319.

Latta, RG, Linhart, YB, Lundquist, L, Snyder, MA (2000). Patterns of monoterpene variation within individual trees in Ponderosa pine. J. Chem. Ecol. 26:1341-1357.

Litvak, ME, Monson, RK (1998). Patterns of induced and constitutive monoterpene production in conifer needles in relation to insect herbivory. Oecologia 114:531-540.

Manninen, A-M, Tarhanen, S, Vuorinen, M, Kainulainen, P (2002). Comparing the variation of needle and wood terpenoids in Scots pine provenances. J. Chem. Ecol. 28:211-228.

Meiners, T, Wäckers, F, Lewis, W (2003). Associative learning of complex odours in parasitoid host location. Chem. Senses 28:231-236.

Meiners, T, Westerhaus, C, Hilker, M (2000). Specificity of chemical cues used by a specialist egg parasitoid during host location. Entomol. Exp. Appl. 95:151-159.

Moore, GE, Clark, EW (1968). Suppressing microorganisms and maintaining turgidity in coniferous foliage used to rear insects in the laboratory. J. Econ. Entomol. 61:1030-1031.

Mumm, R, Schrank, K, Wegener, R, Schulz, S, Hilker, M (2003). Chemical analysis of volatiles emitted by *Pinus sylvestris* after induction by insect oviposition. J. Chem. Ecol. 29:1235-1252.

Petrakis, PV, Tsitsimpikou, C, Tzakou, O, Couladis, M, Vagias, C, Roussis, V (2001). Needle volatiles from five *Pinus* species growing in Greece. Flavour Frag. J. 16:249-252.

Pettersson, J (1970). An aphid sex attractant. I. Biological studies. Entomol. Scand. 1:63-73.

Pschorn-Walcher, H (1988). Die Parasitenkomplexe europäischer Diprionidae in ökologisch-evolutionsbiologischer Sicht. Z. Zool. Syst. Evol. 26:89-103.

Pschorn-Walcher, H (1982). Symphyta. In: Schwenke, W (ed), Die Forstschädlinge Europas Vol. 4. Parey, Hamburg, pp. 66-128.

Pschorn-Walcher, H, Eichhorn, O (1973). Studies on the biology and ecology of the egg parasites (Hym. Chalcidoidea) of the pine sawfly *Neodiprion sertifer* (Geoff.) (Hym.: Diprionidae) in Central Europe. Z. ang. Ent. 74:286-318.

Rafii, ZA, Dodd, RS, Zavarin, E (1996). Genetic diversity in foliar terpenoids among natural populations of European black pine. Biochem. Syst. Ecol. 24:325-339.

Rutledge, CE (1996). A survey of identified kairomones and synomones used by insect parasitoids to locate and accept their hosts. Chemoecology 7:121-131.

Schütt, P, Schuck, HJ, Lang, UM (1992). Handbuch und Atlas der Dendrologie. Ecomed, Landsberg.

Sjödin, K, Persson, M, Fäldt, J, Ekberg, I, Borg-Karlsson, A-K (2000). Occurrence and correlations of monoterpene hydrocarbon enantiomers in *Pinus sylvestris* and *Picea abies*. J. Chem. Ecol. 26:1701-1720.

Sjödin, K, Persson, M, Borg-Karlsson, A-K, Norin, T (1996). Enantiomeric compositions of monoterpene hydrocarbons in different tissues of four individuals of *Pinus sylvestris*. Phytochemistry 41:439-445.

Steidle, JLM, van Loon, JJA (2003). Dietary specialization and infochemical use in carnivorous arthropods: testing a concept. Entomol. Exp. Appl. 108:133-148.

Steidle, JLM, van Loon, JJA (2002). Chemoecology of parasitoid and predator oviposition behaviour. In: Hilker, M, Meiners, T (eds), Chemoecology of Insect Eggs and Egg Deposition. Blackwell Publishing, Berlin, Oxford, pp. 291-317.

Stout, MJ, Bostock, RM (1999). Specificity of induced responses to arthropods and pathogens. In: Agrawal, AA, Tuzun, S , Bent, E (eds), Induced Plant Defenses against Pathogens and Herbivores. APS Press, St. Paul, pp. 183-209.

Sznajder, B, Harvey, JA (2003). Second and third trophic level effects of differences in plant species reflect dietary specialisation of herbivores and their endoparasitoids. Entomol. Exp. Appl. 109:73-82.

Takabayashi, J, Dicke, M (1996). Plant-carnivore mutualism through herbivore-induced carnivore attractants. Trends Plant Sci. 1:109-113.

Takabayashi, J, Takahashi, S, Dicke, M, Posthumus, MA (1995). Developmental stage of herbivore *Pseudaletia separata* affects production of herbivore -induced synomone by corn plants. J. Chem. Ecol. 21:273-287.

Takasu, K, Lewis, WJ (2003). Learning of host searching cues by the larval parasitoid *Microplitis croceipes*. Entomol. Exp. Appl. 108:77-86.

Tingey, DT, Turner, DP, Weber, JA (1991). Factors controlling the emissions of monoterpenes and other volatile organic compounds. In: Sharkey, T, Holland, E, Mooney, H (eds), Trace Gas Emissions by Plants. Academic Press, San Diego, pp. 93-119.

Turlings, TCJ, Fritzsche M.E. (1999). Attraction of parasitic wasps by caterpillar-damaged plants. In: Chadwick, DJ, Goode, JA (eds), Insect-plant interactions and induced plant defence. Wiley & Sons, Chichester, pp. 21-32.

Turlings, TCJ, Benrey, B (1998). Effects of plant metabolites on behavior and development of parasitic wasps. Ecoscience 5:1-13.

Turlings, TCJ, Gouinguené, S, Degen, T, Fritzsche-Hoballah, ME (2002). The chemical ecology of plant-caterpillar-parasitoid interactions. In: Tscharntke, T, Hawkins, B (eds), Multitrophic Level Interactions. Cambridge University Press, Cambridge, pp. 148-173.

Turlings, TCJ, Bernasconi, M, Bertossa, R, Bigler, F, Caloz, G, Dorn, S (1998). The induction of volatile emissions in maize by three herbivore species with different feeding habits: possible consequences for their natural enemies. Biol. Control 11:122-129.

Turlings, TCJ, Wäckers, FL, Vet, LEM, Lewis, WJ, Tumlinson, JH (1993). Learning of host-finding cues by hymenopterous parasitoids. In: Papaj, DR, Lewis, AC (eds), Insect Learning. Ecological and Evolutionary Perspectives. Chapman & Hall, New York, pp. 51-78.

Vet, LEM, Dicke, M (1992). Ecology of infochemical use by natural enemies in a tritrophic context. Annu. Rev. Entomol. 37:141-172.

Vet, LEM, Lewis, WJ, Cardé, RT (1995). Parasitoid foraging and learning. In: Cardé, RT, Bell, WJ (eds), Chemical Ecology of Insects 2. Chapman and Hall, London; New York, pp. 65-101.

Vet, LEM, Van Lenterern, JC, Heymans, M, Meelis, E (1983). An airflow olfactometer for measuring olfactory responses of hymenopterous parasitoids and other small insects. Physiol. Entomol. 8:97-106.

Vinson, SB (1998). The general host selection behavior of parasitoid hymenoptera and a comparison of initial strategies utilized by larvaphagous and oophagous species. Biol. Control 11:79-96.

Vinson, SB (1991). Chemical signals used by parasitoids. Redia 74:15-42.

Zivojinovic, D (1954). *Diprion pini* L. History of the gradation and consequences of the defoliation it caused on the Maljen. Plants Protection (Beograd) 24:3-19.

# Chapter 6

# Analysis of Volatiles from Black Pine (*Pinus nigra*): Significance of Wounding and Egg Deposition by a Herbivorous Sawfly

**Abstract.** The composition of headspace volatiles of black pine (*Pinus nigra*) was analysed by coupled gas chromatography – mass spectrometry (GC-MS). It has been shown in a previous study that egg deposition of the sawfly *Diprion pini* on *P. sylvestris* induces a quantitative change of the pine volatile blend. *Chrysonotomyia ruforum*, an egg parasitoid of *D. pini* is known to be attracted by volatiles from egg-carrying *P. sylvestris*, but not by odour from egg-laden *P. nigra* . Therefore, the present study focused on the question whether also *P. nigra* as another host plant of this sawfly responds to egg deposition by change of its volatile blend. The headspace of untreated, egg-carrying, and artificially wounded *P. nigra* twigs were compared. The artificial damage inflicted to the twigs mimicked the damage by the sawfly female prior to egg deposition. 35 mainly terpenoid compounds that were identified in more than 50% of the egg-carrying *P. nigra* twigs could also detected in the headspace of untreated and artificially wounded twigs.

Quantitative differences of the volatile blends of differently treated *P. nigra* twigs were compared by multivariate data analyses. PLS-DA (projection to latent structures-discriminant analysis) revealed that volatile blends of differently treated *P. nigra* differed significantly. When comparing volatiles from egg-carrying and artificially wounded *P. nigra* with respective *P. sylvestris* samples, qualitative and quantitative differences were detected. The differences in volatile composition of *P. nigra* and *P. sylvestris* are discussed with special respect to the egg parasitoid's response to odours of egg-carrying pine twigs.

**Keywords.** egg deposition, headspace, monoterpenes, *Pinus nigra*, *Pinus sylvestris*, Pinaceae, sawfly, sesquiterpenes, terpenoids, volatiles.

**Introduction**

Like all plants, coniferous trees of the genus *Pinus* are under a continuous threat by numerous herbivorous insects and pathogens attacking nearly all parts and tissues of the plant. To defend themselves, pines produce large amounts of oleoresin that is accumulated in a highly developed network of specialized resin ducts, which are distributed in the wood, bark, and needles (Gijzen et al. 1993; Trapp and Croteau 2001). Compared to other conifer taxa, pines produce and accumulate large amounts of oleoresin constitutively (Gijzen et al. 1993; Lewinsohn et al. 1991; Trapp and Croteau 2001). In addition, it has been shown that defence reactions in pines can be induced by attacking herbivores or pathogens (Bonello et al. 2001; Krokene et al. 2000; Litvak and Monson 1998; Popp et al. 1995; Raffa 1991; Trewhalla et al. 1997). Pine resin is composed of a complex blend of terpenoids consisting of a volatile turpentine fraction [monoterpenes ($C_{10}$), sesquiterpenes ($C_{15}$)], and a less volatile rosin fraction [diterpene resin acids ($C_{20}$)], and their derivatives (Gershenzon and Croteau 1991; Trapp and Croteau 2001).

The volatile fraction of terpenoids can be toxic or repellent to herbivores, defending pines directly against these attacking herbivores (Gijzen et al. 1993, and references therein). Moreover, pine volatiles are also involved in attracting predatory and parasitic insects, thus defending pines indirectly against herbivores (Erbilgin and Raffa 2001; Hilker et al. 2002a; Sullivan and Berisford 2004; Sullivan et al. 2000). Hilker et al. (2002a) showed that volatiles from Scots pine (*Pinus sylvestris* L.) induced by egg deposition of the herbivorous pine sawfly *Diprion pini* L. (Hymenoptera, Diprionidae) attracted the egg parasitoid *Chrysonotomyia ruforum* Krausse (Hymenoptera, Eulophidae). Volatiles from artificially wounded pines without eggs were not attractive to the egg parasitoids. Twigs were artificially wounded to mimic the mechanical damage the sawfly female inflicts with her ovipositor during the egg deposition. Analysis of the volatiles of oviposition-induced twigs of *P. sylvestris* revealed no significant qualitative, but quantitative differences in the volatile terpenoid pattern compared to artificially wounded control twigs (Mumm et al. 2003).

We hypothesized in a previous study (see Chapter 5) that parasitoids would also be attracted by volatiles from other pine species than *P. sylvestris* after egg deposition of *D. pini*. Several studies showed that black pine, *P. nigra,* is readily accepted as a host by *D. pini* for oviposition, even though egg development and larval performance is significantly reduced on this pine species (Auger et al. 1994; Barre et al. 2002; Eliescu 1932; Zivojinovic 1954). Therefore, the egg parasitoid's response to egg-carrying black pine was tested by Mumm (2004). However, volatiles from *P. nigra* twigs on which *D. pini* had deposited eggs were not attractive to *C. ruforum* females.

In the present study, we conducted a detailed GC-MS analysis of volatiles emitted by twigs of *P. nigra* (var. *nigra*). We analysed how egg deposition by *D. pini* or artificial wounding affects the volatile pattern emitted by *P. nigra* twigs. Though several studies analysed the chemical composition of essential oils from *P. nigra* needle tissue (e.g. Macchioni et al. 2003; Rafii et al. 1996; Rezzi et al. 2001; Roussis et al. 1995), far less is known how wounding, either artificial wounding or damage due to herbivores, affects the composition of volatiles emitted by *P. nigra.*

In addition to these "within-species" headspace comparisons of differently treated *P. nigra,* also volatile blends from *P. nigra* and *P. sylvestris* were compared. These "between-species" headspace comparisons were conducted in order to provide further information on the quality and quantity of volatiles able to attract the egg parasitoid *C. ruforum* to oviposition-induced pine. We statistically compared the volatile compositions of artificially wounded and egg-carrying twigs of *P. nigra* (non-attractive to *C. ruforum*) with the volatile composition of oviposition-induced (attractive) and artificially wounded (non-attractive) twigs of *P. sylvestris.* The data set of *P. sylvestris* volatiles used for the comparison was collected in a previous study (Mumm et al. 2003).

**Material and methods**

*Plants and insects*

*P. nigra* is naturally distributed in Southern Europe ranging from Spain, Austria, Balkan Peninsula, Turkey up to the Black Sea, but has been cultivated for parks and forests worldwide (Krüssmann 1983; Rafii et al. 1996; Schütt et al. 1992). Depending on the geographical distribution, *P. nigra* is separated into 4 - 5 subspecies and several varieties (Schütt et al. 1992, and references therein). The subspecies *P. nigra* (var. *nigra*) investigated here is naturally distributed mainly in south Eastern Europe (Austria, Italy, Balkan Peninsula). Branches of *P. nigra var. nigra* were detached from ca. 14-year-old trees growing in a pine stand near Kremmen (Brandenburg, Germany). The lower part of a branch was cleaned and sterilized according to Moore and Clark (1968). Branches were kept in water at 10 °C, L18 : D6, and 70% relative humidity in a climate chamber until treatment. The pine sawfly *D. pini* was reared on twigs of *P. sylvestris* L. in the laboratory as described by Bombosch and Ramakers (1976) and Eichhorn (1976) at 25 °C, L18 : D6 and 70% rh.

*Plant treatments general*

One major interest of this study was to investigate whether egg deposition or artificially wounding induces changes in the volatile pattern of *P. nigra*. Therefore, small twigs of approx. 12 cm length were cut from a branch, provided with water and treated for a period of 72 h at 25 °C, L18 : D6 and 70% rh as described below. During this treatment period, all pine twigs were placed in a separate glass jar (height: 22 cm, 15 cm i.d.) covered by a gauze lid. Only one sample per tree was taken for experiments. For treatments, test pine twigs were either exposed to sawflies or were artificially wounded. For control, untreated pine twigs were cut and kept like treated twigs. Egg deposition by *D. pini* on cut *P. sylvestris* twigs has previously been shown to induce changes in the volatile pattern that attract the egg parasitoid *C. ruforum* (Hilker et. al. 2002a; Mumm et al. 2003). Headspace samples were taken from control twigs and from test twigs immediately after treatment (see below).

*Pine twig with eggs of D. pini*

Two males and females of *D. pini* were offered a pine twig for egg deposition. When *D. pini* females had laid at least four egg masses after 72 h, volatiles of the pine headspace were collected.

*Artificially wounded and untreated pine twigs*

Pine twigs were artificially damaged by slitting eight needles longitudinally with a scalpel to mimic the mechanical damage that a *D. pini* female inflicts with her ovipositor prior to oviposition (Hilker et al. 2002a).

*Collection of headspace volatiles*

Volatile compounds emitted by the pine twigs were collected using a dynamic headspace sampling method as described by Mumm et al. (2003). The cut end of a twig was tightly wrapped with Parafilm® during the sampling period of 5h. Volatile collection started 6h after beginning of the photophase. The sample conditions were the same as those used for the treatments (see above). Air was purified by passing activated charcoal before entering the flasks (250 ml) containing the plant samples. Volatiles emitted by the twigs were collected on 5 mg charcoal filters (Precision charcoal filter, K. Trott, Germany) and were eluted with 50 µl dichloromethane containing 12.5 ng / µl methyl octanoate as internal standard. Flowmeters (Supelco, Germany) ensured a constant airflow of 110 ml / min. All components of the sampling set-up were connected with teflon tubings.

*Chemical analysis of headspace samples*

Headspace samples were analysed by GC-MS. For identification of components, samples were analysed using a Hewlett Packard GC 6890 MSD 5973 with a split/splitless inlet, equipped with a 30 m HP5MS capillary column (id = 0.25 mm, df = 0.25 µm). The oven temperature was programmed from 50 °C (5 min hold) to 300 °C at 5 °C/min. Helium flow was adjusted to 1 ml/min in a constant flow mode. The mass spectrometer was operated in 70 eV EI ionization mode. Compounds were identified by comparing mass spectra and retention indices with those of reference compounds or mass spectral library data (MassFinder 2.2, NIST library). Retention indices were calculated for each compound according to van den Dool and Kratz (1963) and compared to critically evaluated, tabulated data (Adams 1989).

For quantification, analysis was conducted by use of a Fisons GC model 8060 and a Fisons MD 800 quadrupole MS using a DB5-MS capillary column (30 m, 0.32 mm i.d., film thickness 0.25 μm, J&W, USA). The samples (1 μl) were injected in splitless mode (injector temperature 240 °C) with helium as carrier gas (inlet pressure 10 kPa). The temperature program started at 40 °C (4 min hold) and rose with 10 °C/min up to 280 °C. The column effluent was ionised by electron impact ionisation (EI) at 70 eV. Mass range was from 35 - 350 *m/z* with a scan time of 0.9 s and an interscan delay of 0.1 s. Headspace samples were taken from 10 *P. nigra* twigs carrying eggs of *D. pini*, 10 artificially wounded twigs, as well as from 20 twigs that were left untreated (controls).

*Statistical analysis*

For quantitative analysis, the 35 most abundant volatile compounds (i.e. those components detected in more than 50% of the egg-carrying *P. nigra* twigs) were selected. We excluded three components prior to the analysis because it has been shown that (*E*)-β-ocimene can partly be decomposed into (*E,E*)-2,6-dimethyl-1,3,5,7-octatetraene and (*E,E*)-2,6-dimethyl-3,5,7-octatrien-2-ol, a process that is catalysed by activated charcoal, as it was used as adsorbent (W. Boland, Jena, personal communication). Thus, the amounts of these components might be artefacts.

Relative amounts were evaluated by multivariate data analysis using the software programme SIMCA-P 10.5 (Umetrics AB, Umeå, Sweden). The data were subjected to principal components analysis (PCA) and PLS-DA (projection to latent structures-discriminant analysis) (Wold et al. 1989). PCA was conducted in order to extract and display the systematic variation in the multivariate data set consisting of the quantities of 35 compounds (variables) (see below) obtained from 10 replicates (observations). In PCA the multivariate data set is projected down to a lower-dimensional plane formed by the principal components (PCs) which approximate the data as well as possible in the least square sense (Eriksson et al. 2001; Jackson 1991).

By projecting all observations on this low-dimensional plane, so called scores are obtained that visualize the structure of the investigated data in a score plot (Eriksson et al. 2001). The corresponding loading plot displays the relationships among the variables, i.e. which variables are important and which are not, and how the important variables combine to separate the clusters of observations shown in the score plot (Eriksson et al. 2001). Important variables are located on the periphery of the loading plot, unimportant variables are encountered around the origin of the plot (0,0). Score and loading plot are complementary and super-imposable, i.e. an interesting pattern seen in the score plot can be interpreted by looking along that interesting direction in the loading plot.

Raw data (integrated peak areas) were normalized, i.e. peak areas of the 35 compounds (**X** variables) were summed to 100 and percentage of each variable was calculated (Tab. 1). A selective normalisation of the raw data to avoid the effect of closure as suggested by Johansson et al. (1984) was not necessary, since major components were not significantly negatively correlated. The normalized data were transformed to log (X+ 0.00001). The constant 0.00001 was added to provide non-detectable components with a small non zero value (Sjödin et al. 1989). Transformed variables were then mean-centered and scaled to unit variance and represented as a matrix **X**.

In PLS-DA the data set is modelled in way similar to PCA, but in combination with a discriminant analysis. The objective of PLS-DA is to find a model that separates (discriminates) the **X** data according to the above described treatments as good as possible (Eriksson et al. 2001). Therefore, an additional **Y** matrix was made up as a dummy variable, containing the values 1 and 0 for each treatment, respectively. The number of significant principal components was determined by cross-validation (Wold et al. 1989; Eriksson et al. 2001). The ellipse shown in score plots defines the Hotelling's $T^2$ confidence region (95%). Observations projected outside the ellipse of the model were detected as outlier and were discarded from the model (Eriksson et al. 2001).

The data set of *P. sylvestris* volatiles used for the comparison with the volatile pattern of *P. nigra* was obtained from a previous study conducted by Mumm et al. (2003). Relative amounts of components were normalized and transformed as described for the data of *P. nigra*. Transformed variables of both pine species were then mean-centred and scaled to unit variance.

## Results and discussion

### Analysis of Pinus nigra volatiles

A list of the most abundant 35 compounds that were detected in more than 50% of the samples taken from egg-carrying twigs of P. nigra twigs is presented in Table 1 The majority of compounds is of terpenoid origin, i.e. monoterpenes, sesquiterpenes, or derivatives of these. In addition, (Z)-3-hexen-1-ol (**1**) was identified. (Z)-3-hexen-1-ol is rather uncommon in conifers, whereas it is well-known for angiosperm plants (Hatanaka et al. 1995; Paré and Tumlinson 1999). Many terpenoid compounds are typical constituents of conifer resin and have been reported from *P. nigra* foliage before (Tab. 1) (e.g. Bojovic 1997, Holzke 2001; Macchioni et al. 2003; Rafii et al. 1996; Rezzi et al. 2001; Roussis et al. 1995). Regardless of how the pine twigs were treated, α-pinene (**3**), β-pinene (**6**), β-myrcene (**7**), limonene (**10**), and β-phellandrene (**11**) were the major monoterpenes. The components (E)-β-caryophyllene (**27**) and germacrene D (**32**) were the predominating sesquiterpenes confirming former studies (Macchioni et al. 2003; Rafii et al. 1996; Rezzi et al. 2001, Roussis et al. 1995, and references therein). Macchioni et al. (2003) analysed essential oils from needles, branches without needles, and cones of *P. nigra* separately. They detected the components 1,8-cineole (**12**), pinocamphone (**18**), myrtenal (**20**) in branches and cones only, but not in needle tissue. On the other hand, the components α-terpinyl acetate (**24**), β-cubebene (**25**), (E)-β-farnesene (**28**), and α-humulene (**29**) were exclusively detected in needle tissue by Macchioni et al. (2003).

**Table 1** Mean percentages of relative amounts based on total integrated peak area of 35 volatile compounds after different treatments are given. Coefficients of variation are shown in parentheses. Volatiles of 10 samples of *P. nigra* twigs after egg deposition and wounding, respectively, and 20 untreated samples were analysed. [a]Percentages of samples in which the compounds have been detected. *RI:* Retention index was calculated for each compound according to van den Dool and Kratz (1963). Compounds marked with an asterisk correspond to compounds identified in *P. sylvestris* after similar treatments by Mumm et al. (2003). These compounds were included for comparative analysis of *P. nigra* and *P. sylvestris* volatiles.

| No. | RI | Compound | Egg deposition | | | Wounding | | | No Treatment | | |
|---|---|---|---|---|---|---|---|---|---|---|---|
| 1 | 851 | (Z)-3-Hexen-1-ol | 0.54 | (0.44) | 80[a] | 0.72 | (0.59) | 90[a] | 0.23 | (0.39) | 50[a] |
| 2 | 926 | 3-Thujene* | 0.42 | (1.03) | 80 | 0.38 | (0.75) | 90 | 0.55 | (0.77) | 85 |
| 3 | 932 | α-Pinene* | 27.75 | (0.62) | 100 | 26.81 | (0.83) | 100 | 44.55 | (0.67) | 100 |
| 4 | 946 | Camphene* | 0.79 | (0.66) | 100 | 0.76 | (1.12) | 100 | 0.83 | (1.18) | 100 |
| 5/6 | 974 | Sabinene/β-Pinene* | 10.95 | (0.76) | 100 | 7.42 | (0.94) | 100 | 4.79 | (0.76) | 100 |
| 7 | 991 | β-Myrcene* | 5.17 | (1.18) | 100 | 7.05 | (1.63) | 100 | 5.08 | (0.75) | 100 |
| 8 | 1003 | α-Phellandrene | 0.87 | (0.52) | 90 | 0.10 | (0.25) | 70 | 0.41 | (0.46) | 90 |
| 9 | 1008 | 3-Carene* | 3.83 | (2.87) | 80 | 1.17 | (1.25) | 100 | 1.56 | (1.19) | 100 |
| 10 | 1028 | Limonene* | 9.69 | (0.64) | 100 | 24.81 | (0.93) | 100 | 8.46 | (0.77) | 100 |
| 11 | 1028 | β-Phellandrene* | 5.98 | (1.37) | 70 | 3.42 | (1.05) | 40 | 9.32 | (0.67) | 85 |
| 12 | 1028 | 1,8-Cineole | 0.50 | (1.60) | 100 | 2.14 | (0.85) | 100 | 0.26 | (1.67) | 95 |
| 13 | 1059 | γ-Terpinene* | 0.12 | (1.71) | 60 | 0.42 | (0.73) | 80 | 0.21 | (0.54) | 90 |
| 14/15 | 1088 | Terpinolene/Cymenene* | 0.61 | (1.48) | 90 | 0.44 | (1.53) | 100 | 0.48 | (0.91) | 100 |
| 16 | 1140 | (E)-Pinocarveol | 1.62 | (0.30) | 80 | 0.13 | (0.67) | 50 | 0.17 | (0.46) | 85 |
| 17 | 1146 | (E)-Verbenol | 0.41 | (1.05) | 80 | 0.05 | (1.20) | 30 | 0.28 | (0.67) | 90 |
| 18 | 1163 | Pinocamphone | 0.65 | (0.67) | 90 | 0.04 | (1.77) | 30 | 0.19 | (0.90) | 95 |
| 19 | 1167 | Borneol | 1.09 | (1.23) | 80 | 0.23 | (1.21) | 70 | 0.14 | (2.57) | 95 |
| 20 | 1202 | Myrtenal | 0.89 | (0.83) | 70 | 0.01 | (1.58) | 10 | 0.11 | (1.54) | 65 |
| 21 | 1235 | Thymyl methyl ether* | 0.17 | (0.75) | 80 | 0.17 | (0.93) | 40 | 0.09 | (2.41) | 70 |
| 22 | 1256 | Linalyl acetate | 0.33 | (0.84) | 90 | 0.33 | (1.59) | 20 | 0.04 | (0.76) | 50 |
| 23 | 1287 | Bornyl acetate* | 1.27 | (1.13) | 100 | 2.05 | (1.62) | 100 | 0.94 | (0.78) | 100 |
| 24 | 1351 | α-Terpinyl acetate | 0.81 | (0.80) | 90 | 0.30 | (1.24) | 50 | 0.17 | (1.74) | 90 |
| 25/26 | 1392 | β-Cubebene/β-Elemene* | 0.34 | (1.01) | 90 | 0.26 | (1.43) | 30 | 1.15 | (0.85) | 100 |
| 27 | 1428 | (E)-β-Caryophyllene* | 10.78 | (0.93) | 100 | 8.84 | (1.06) | 100 | 5.69 | (1.38) | 100 |
| 28 | 1457 | (E)-β-farnesene* | 0.30 | (1.22) | 100 | 1.73 | (2.77) | 100 | 1.30 | (1.57) | 95 |
| 29 | 1462 | α-Humulene* | 1.43 | (1.51) | 100 | 1.08 | (3.16) | 100 | 0.95 | (1.11) | 100 |
| 30 | 1478 | (Z)-Muurola4(15),5-diene | 0.46 | (0.65) | 90 | 0.08 | (1.63) | 60 | 0.77 | (1.32) | 75 |
| 31 | 1483 | γ-Muurolene | 0.59 | (0.84) | 80 | 0.49 | (0.76) | 90 | 1.69 | (1.22) | 85 |
| 32 | 1490 | Germacrene D* | 10.52 | (0.86) | 100 | 7.66 | (1.86) | 100 | 8.89 | (1.23) | 95 |
| 33 | 1514 | α-Muurolene* | 0.29 | (0.85) | 60 | 0.32 | (1.98) | 70 | 0.05 | (1.35) | 5 |
| 34 | 1522 | γ-Cadinene* | 0.39 | (1.45) | 90 | 0.19 | (2.10) | 50 | 0.28 | (1.20) | 95 |
| 35 | 1530 | δ-Cadinene* | 0.45 | (1.20) | 80 | 0.38 | (1.29) | 70 | 0.36 | (4.47) | 95 |

Several studies showed that the enantiomeres of chiral monoterpenes should be considered as single components which may provide additional information for a chemical discrimination of conifer species, chemotypes or tissues (Cool and Zavarin 1992; Croteau 1987; Sjödin et al. 1996, 2000; Valterová et al. 1995). However, in response to wounding or herbivory the changes in the enantiomeric composition are weak (Fäldt et al. 2001; Sadof and Grant 1997; Sjödin ct al. 1993). Egg deposition of *D pini* was shown to induce no changes of the enantiomeric composition of monoterpenes in *P. sylvestris* (Mumm et al. 2003). Thus, we dispensed with an analysis of the enantiomeres, since no essential information for a better resolution of volatile patterns of differently treated *P. nigra* could be expected from it.

To study whether egg deposition by *D. pini* affects the composition of *P. nigra* headspace, volatiles from egg-carrying pine twigs were compared to the headspace of artificially wounded twigs and untreated controls. All volatile components detected in more than 50% of the samples of *P. nigra* twigs carrying eggs from *D. pini* were also emitted by artificially wounded and undamaged control twigs. This result is consistent with the findings in *P. sylvestris* which show that all volatiles analysed from egg-carrying *P. sylvestris* twigs are also present in the headspace of untreated and artificially wounded twigs (Mumm et al. 2003). However, the three sample types of *P. nigra* released the 35 components in different frequencies. For example, myrtenal (**20**) and linalyl acetate (**22**) were detected only occasionally in the headspace of artificially wounded *P. nigra* twigs, but were frequently present in the headspace of egg-carrying and untreated twigs (Tab. 1). On the other hand, $\alpha$-muurolene (**33**) was only detected in 5% of the untreated pine samples, but was identified in 60% of twigs with eggs and 70% of artificially wounded twigs, respectively (Tab. 1).

To compare the volatile patterns of the differently treated *P. nigra* samples with respect to the quantities of volatile compounds, a PCA was conducted including the relative amounts of the 35 volatiles. This PCA resulted in a model with two significant principal factors, explaining 35% of the variance in the data. PC-score plots show a summary of the relationships among the observations.

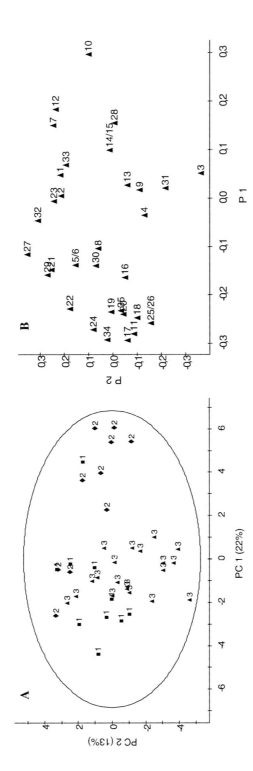

**Figure 1** Analysis of the volatile pattern of differently treated *P. nigra*. Score (A) plot and loading plot (B) from principal component analysis (PCA) based on relative amounts of 35 volatile constituents of *Pinus nigra*. A total of 35 % of the variance in the data is explained by the two significant principal components, as judged by cross-validation. The ellipse shown in the score plot defines the Hotelling's $T^2$ confidence region (95 %). Box (1) = twigs carrying eggs of *D. pini*, Diamond (2) = artificially wounded twigs, Triangle (3) = untreated twigs. Numbers in the loading plot symbolize compounds given in Table 1.

The score plot of the *P. nigra* data indicates that volatile blends of the three sample types [(1) egg-carrying, (2) artificially damaged, (3) untreated] are dissimilar by forming three groups, though some overlapping occurs (Fig. 1A). The majority of artificially damaged *P. nigra* samples form a separate group on the right of the plot (Fig. 1A). These artificially wounded samples are characterized especially by higher amounts of limonene (**10**), β-myrcene (**7**), 1,8-cineole (**12**), and (*E*)-β-farnesene (**28**) which are located also in the right part of the corresponding loading plot (Fig. 1B). Only one of the *P. nigra* samples carrying eggs was projected within the cluster of the artificially damaged twigs (Fig. 1A). The second principal factor, explaining 13% of the variance, tended to separate untreated twig samples from the others (Fig. 1A). Volatile blends of untreated pines are characterized by high amounts of α-pinene (**3**) and low amounts of (*E*)-β-caryophyllene (**27**) and α-humulene (**29**) compared to egg-carrying or artificially wounded pine twigs (Fig 1B). Thus, the PCA indicates that artificial wounding of *P. nigra* twigs resulted in a more pronounced change of the volatile pattern than egg deposition when compared to untreated controls (Fig.1A, B).

The difference in composition of *P. nigra* volatile blends collected from the three sample types [(1) egg-carrying, (2) artificially damaged, (3) untreated], as indicated by the PCA, was found to be statistically significant as determined by the PLS-DA (Fig. 2A). Two significant discriminant factors were found, explaining 33% of the variance of **X** variables and 54% of the variation of the dummy variables **Y**. Three groups according to the treatments were fairly discriminated. Nearly all artificially wounded samples were clearly separated from the other two treatments by the first discriminant factor explaining 20% of the variance. The second discriminant factor explaining 13% of the variance separates the untreated samples from the egg-laden ones. However, the discrimination between the volatile composition of untreated and egg-carrying pine twigs again revealed to be less pronounced (Fig. 2A).

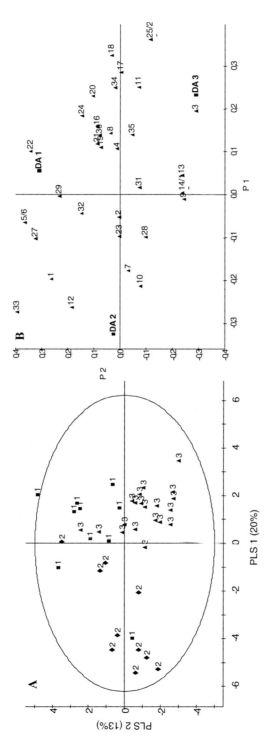

**Figure 2** Analysis of the volatile pattern of differently treated *P. nigra*. Score plot (A) and loading plot (B) from PLS-DA based on relative amounts of volatile constituents of *P. nigra*. A total of 33 % of the variance of the X matrix and 54 % of the variance of the dummy variable Y is explained by the two significant principal components, as judged by cross-validation. The ellipse shown in the score plot defines the Hotelling's $T^2$ confidence region (95 %). Box (1) = twigs carrying eggs of *D. pini*, Diamond (2) = artificially wounded twigs, Triangle (3) = untreated twigs. Numbers in the plot symbolize compounds given in Table 1. DA1, DA2 and DA3 denote the dummy variables Y that show the typical triangular distribution because of their 0/1-nature.

The PLS-DA loading plot given in Figure 2B shows which variables contribute strongly to the separation of classes. The most important components for resolving artificially damaged samples are limonene (**10**), 1,8-cineole (**12**), and α-muurolene (**33**). α-Pinene (**3**) is the most important compound for resolving untreated *P. nigra* samples, while linalyl acetate (**22**) and α-humulene (**29**) are important for resolving volatile samples of *P. nigra* carrying eggs of *D. pini* (Fig. 2B).

The models presented here reveal significant quantitative changes of the volatile composition of *P. nigra* twigs due to wounding and insect egg deposition. The variation in the composition of volatile secondary constituents of conifers has widely been shown to be very high (e.g. Latta et al. 2003; Manninen et al. 2002; Petrakis et al. 2001). Thus, our models only explain a part (33-35%) of the variance in the data, i.e. only a part of the variation found in the data is due to differences between treated and untreated twigs (Sjödin et al. 1989).

Our analysis shows that the egg deposition by the female sawfly obviously induces changes in the volatile blend of *P. nigra* that is different from the blend emitted after artificial wounding. Many studies demonstrated that wounding could not completely simulate the effects of herbivore damage (e.g. Baldwin 1990; de Bruxelles and Roberts 2001; Litvak and Monson 1998). Such differences can often be explained by the involvement of specific herbivore-derived elicitors (recent reviews by Baldwin et al. 2001; Dicke and van Poecke 2002; Hilker et al. 2002b). An elicitor associated with oviposition of *D. pini* has been found to be located in the oviduct secretion coating the eggs. Application of oviduct secretion into artificially wounded needles tissue of *P. sylvestris* induces the emission of volatiles attractive to the egg parasitoid, whereas artificially wounded needles without elicitor treatment did not emit an attractive volatile blend (Hilker et al. 2002a). Numerous studies have shown that damage of plant tissue caused by herbivores induces qualitatively different or quantitatively more pronounced responses in plants than just artificial damage (Dicke 1999, and references therein).

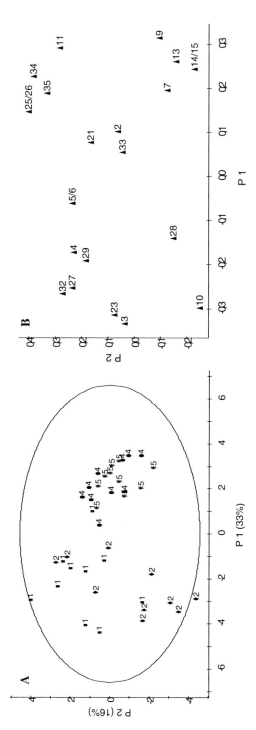

**Figure 3** Comparison of volatile pattern of *P. nigra* and *P. sylvestris*. Score plot (A) and loading plot (B) from principal component analysis (PCA) based on relative amounts of 20 corresponding volatile constituents of *Pinus nigra* and *P. sylvestris*. A total of 49 % of the variance in the data is explained by the two significant principal components, as judged by cross-validation. The ellipse defines the Hotelling's $T^2$ confidence region (95 %). Box (1) = twigs of *P. sylvestris*, Diamond (2) = artificially wounded twigs of *P. nigra*, Dots (4) = twigs of *P. sylvestris* carrying eggs of *D. pini*, Stars (5) = artificially wounded twigs of *P. sylvestris*; Numbers in the plot indicate compounds that correspond to the volatile pattern of both pine species. These compounds are marked with an asterisk in Table 1.

However, there is some evidence that herbivores or the application of the eliciting secretions may suppress specific wounding-induced responses in plants (Felton and Eichenseer 1999; Kahl et al. 2000; Musser et al. 2002). Our results also show that the volatile blend of artificially wounded pine differs stronger from untreated pine volatiles than the volatiles released after egg deposition. These findings may suggest that the artificial damage mimicking the damage inflicted by oviposition induces a strong change of volatiles which is attenuated when eggs and oviduct secretion are laid into the oviposition wound.

*Comparison of the volatile composition of P. nigra and P. sylvestris*

We compared the composition of volatiles released by artificially wounded and egg carrying twigs of *P. nigra* with that of egg-carrying (oviposition-induced) and artificially wounded twigs of *P. sylvestris*. For both species we quantitatively analysed only components that were detected in more than 50% of pine samples on which sawfly eggs had been deposited. We found 20 compounds that were present in the headspace of both *P. nigra* and *P. sylvestris*. These components are marked with an asterisk in Table 1. However, another 20 compounds did not match, i.e. they were only detected in either one of the pine species. Thus, there are clear qualitative differences in the blends of volatile of both pine species. Chemical compounds that are exclusively released by *P. sylvestris* might of course be essential cues used by the egg parasitoid *C. ruforum* for host location. Nevertheless, we compared by means of multivariate data analysis the quantitative composition of the 20 components detected both in *P. nigra* and *P. sylvestris*.

A PCA based on the relative amounts of the 20 components detected in *P. nigra* and *P. sylvestris* resulted in a model with two significant principal factors explaining 49% of the variance. Two separate groups corresponding to the two pine species were formed in the model (Fig. 3A). The volatile composition of *P. nigra* (denoted by 1 and 2) was clearly distinguishable from the composition of *P. sylvestris* (denoted by 4 and 5). Compared to *P. sylvestris,* the headspace of *P. nigra* contained a higher proportion of α-pinene (**3**), limonene (**10**), bornyl acetate (**23**), and germacrene D (**32**), and lower proportions of 3-carene (**9**), β-phellandrene (**11**), and α-terpinene (**13**) (Fig. 3B).


**Conclusions**


Wounding and insect egg deposition were shown to significantly affect the quantitative composition of the headspace of *P. nigra*. Furthermore, the volatile blend of egg-laden *P. nigra* differs from the odour of egg-carrying *P. sylvestris* both quantitatively and qualitatively.


The positive response of egg parasitoids to oviposition-induced *P. sylvestris* volatiles and their "non-response" to egg-carrying *P. nigra* twigs might be due to (a) the qualitative differences between the volatile components in the headspace of the two pine species (compare Tab. 1), or (b) the quantitative differences between components present in both species as shown in Fig. 3. Regarding qualitative differences, there might be key compounds in the volatile blend of *P. sylvestris* that are *per se* attractive to egg parasitoids or are essential components of an attractive blend. Such key components might just be missing in the headspace of egg-carrying *P. nigra*. Conversely, compounds that are additionally emitted by *P. nigra* may mask the attractiveness of key compounds in the volatile mixture, a phenomenon also referred to as mixture suppression (Chandra and Smith 1998, Laloi et al. 2000, Meiners et al. 2003, and references therein).


However, also the different quantitative composition of egg-laden *P. nigra* could be responsible for the "non-response" of the egg parasitoids to this pine species. Mumm et al. (2003) detected only quantitative differences in the blend of oviposition-induced (attractive) and artificially damaged (not attractive) *P. sylvestris*. Especially the sesquiterpene (*E*)-β-farnesene was emitted in significantly higher amounts in oviposition-induced twigs compared to artificially-wounded *P. sylvestris* twigs (Mumm et al. 2003). Thus, an increased ratio of (*E*)-β-farnesene to other pine volatiles might be used by the egg parasitoid *C. ruforum* to locate pine twigs with host eggs. In contrast, the proportion of (*E*)-β-farnesene was lower in *P. nigra* twigs carrying eggs than in wounded twigs (Fig. 1B). In order to elucidate which quality and quantities of volatiles the egg parasitoid *C. ruforum* needs to recognise an oviposition-induced pine, future studies will address (*inter alia*) the question whether the parasitoid shows a positive behavioural response especially to (*E*)-β-farnesene or to different portions of (*E*)-β-farnesene within a blend of pine volatiles.

## Acknowledgements

Many thanks are due to Frank Müller for his assistance in volatile collections and GC-MS-analyses. We are also very grateful to Ute Braun who helped to rear *Diprion pini*. Dr. Ruther gave helpful comments on earlier drafts of the manuscript. This study was supported by the Deutsche Forschungsgemeinschaft (DFG Hi 416/11-1,2).

## References

Adams, RP (1995). Identification of essential oil - components by gas chromatography/mass spectroscopy. Allured Publishing Corporation, Carol Stream, Illinois USA.

Auger, M-A, Géri, C, Allais, J-P (1994). Effect of the foliage of different pine species on the development and on the oviposition of the pine sawfly *Diprion pini* L. (Hym., Diprionidae); II. Influence on egg laying and interspecific variability of some active secondary compounds. J. Appl. Ent. 117:165-181.

Baldwin, IT (1990). Herbivory simulations in ecological research. Trends Ecol. Evol. 5:91-93.

Baldwin, IT, Halitschke, R, Kessler, A, Schittko, U (2001). Merging molecular and ecological approaches in plant-insect interactions. Curr. Opin. Plant Biol. 4:351-358.

Barre, F, Milsant, F, Palasse, C, Prigent, V, Goussard, F, Géri, C (2002). Preference and performance of the sawfly *Diprion pini* on host and non-host plants of the genus *Pinus*. Entomol. Exp. Appl. 102:229-237.

Bojovic, S (1997). Une approche de la taxonomie du pin noir par la synthèse des caractères terpéniques et morphologiques. Bull. Soc. linn. Provence 48:147-156.

Bombosch, S, Ramakers, PMJ (1976). Zur Dauerzucht von *Gilpinia hercyniae* Htg. Z. Pfl. Krankh. 83:40-44.

Bonello, P, Gordon, TR, Storer, AJ (2001). Systemic induced resistance in Monterey pine. For. Path. 31:99-106.

Chandra, S, Smith, BH (1998). An analysis of synthetic processing of odor mixtures in the honeybee (*Apis mellifera*). J. Exp. Biol. 201:3113-3121.

Cool, LG, Zavarin, E (1992). Terpene variability of mainland *Pinus radiata*. Biochem. Syst. Ecol. 20:133-144.

Croteau, R (1987). Biosynthesis and catabolism of monoterpenoids. Chem. Rev. 87:929-954.

de Bruxelles, GL, Roberts, MR (2001). Signals regulating multiple responses to wounding and herbivores. Crit. Rev. Plant Sci. 20:487-521.

Dicke, M (1999). Evolution of induced indirect defense of plants. In: Tollrian, R, Harvell, CD (eds), The Ecology and Evolution of Inducible Defenses. Princeton University Press, Princeton, pp. 62-88.

Dicke, M, van Poecke, RMP (2002). Signalling in plant-insect interactions: signal transduction in direct and indirect plant defence. In: Scheel, D, Wasternack, C (eds), Plant Signal Transduction. Oxford University Press, Oxford, pp. 289-316.

Eichhorn, O (1976). Dauerzucht von *Diprion pini* L. (Hym.: Diprionidae) im Laboratorium unter Berücksichtigung der Fotoperiode. Anz. Schädlingskd. Pfl. 49:38-41.

Eliescu, G (1932). Beiträge zur Kenntnis der Morphologie, Anatomie und Biologie von *Lophyrus pini*. Z. ang. Ent. 19:22 (188)-67 (206).

Erbilgin, N, Raffa, KF (2001). Modulation of predator attraction to pheromones of two prey species by stereochemistry of plant volatiles. Oecologia 127:444-453.

Eriksson, L, Johansson, E, Kettaneh-Wold, N, Wold, S (2001). Multi- and Megavariate Data Analysis; Principles and Applications. Umetrics Academy, Umeå.

Fäldt, J, Sjödin, K, Persson, M, Valterová, I, Borg-Karlson, A-K (2001). Correlation between selected monoterpene hydrocarbons in the xylem of six *Pinus* (Pinaceae) species. Chemoecology 11:97-106.

Felton, GW, Eichenseer, H (1999). Herbivore saliva and its effects on plant defense against herbivores and pathogens. In: Agrawal, AA, Tuzun, S, Bent, E (eds), Induced Plant Defenses against Pathogens and Herbivores. APS Press, St. Paul, pp. 19-36.

Gershenzon, J, Croteau, R (1991). Terpenoids. In: Rosenthal, GA, Berenbaum, MR (eds), Herbivores Their Interactions with Secondary Metabolites Vol. 1 The Chemical Participants. Academic Press, San Diego, pp. 165-219.

Gijzen, M, Lewinsohn, E, Savage, TJ, Croteau, R (1993). Conifer monoterpenes. In: Teranishi, R, Buttery, RG, Sugisawa, H (eds), Bioactive Volatile Compounds from Plants. ACS Symposium Series 525, Washington, DC, pp. 8-22.

Hatanaka, A, Kajiwara, T, Matsui, K(1995). The biogeneration of green odour by green leaves and it's physiological functions – past, present and future. Z. Naturforsch. 50c:467-472

Hilker, M, Kobs, C, Varama, M, Schrank, K (2002a). Insect egg deposition induces *Pinus* to attract egg parasitoids. J. Exp. Biol. 205:455-461.

Hilker, M, Rohfritsch, O, Meiners, T (2002b). The plant's response towards insect egg deposition. In: Hilker, M, Meiners, T (eds), Chemoecology of Insect Eggs and Egg Deposition. Blackwell Publishing, Berlin, Oxford, pp. 205-233.

Holzke C (2001). Untersuchungen zur Biosynthese und zum Emissionsverhalten ausgesuchter Terpenoide bei Pflanzen. PhD Thesis, University of Cologne.

Johansson, E, Wold, S, Sjödin, K (1984). Minimizing effects of closure on analytical data. Anal. Chem. 56:1685-1688.

Kahl, J, Siemens, DH, Aerts, RJ, Gäbler, R, Kühnemann, F, Preston, CA, Baldwin, IT (2000). Herbivore-induced ethylene suppresses a direct defense but not a putative indirect defense against an adapted herbivore. Planta 210:336-342.

Krokene, P, Solheim, H, Långström, B (2000). Fungal infection and mechanical wounding induce disease resistance in Scots pine. Eur. J. Plant Path. 106:537-541.

Krüssmann, G (1983). Handbuch der Nadelgehölze. Parey, Berlin.

Laloi, D, Bailez, O, Blight, MM, Roger, B, Pham-Delegue, MH (2000). Recognition of complex odours by restrained and free-flying honey bees, *Apis mellifera*. J. Chem. Ecol. 26:2307-2319.

Lewinsohn, E, Gijzen, M, Croteau, R (1991). Defense mechanisms of conifers. Plant Physiol. 96:44-49.

Litvak, ME, Monson, RK (1998). Patterns of induced and constitutive monoterpene production in conifer needles in relation to insect herbivory. Oecologia 114:531-540.

Macchioni, F, Cioni, PL, Flamini, G, Maccioni, S, Ansaldi, M (2003). Chemical composition of essential oil from needles, branches and cones of *Pinus pinea*, *P. halepensis*, *P. pinaster*, and *P. nigra* from central Italy. Flavour Fragr. J. 18:139-143.

Meiners, T, Wäckers, FL, Lewis, WJ (2003). Associative learning of complex odours in parasitoid host location. Chem. Senses 28:231-236.

Moore, GE, Clark, EW (1968). Suppressing microorganisms and maintaining turgidity in coniferous foliage used to rear insects in the laboratory. J. Econ. Ent. 61:1030-1031.

Mumm, R, Schrank, K, Wegener, R, Schulz, S, Hilker, M (2003). Chemical analysis of volatiles emitted by *Pinus sylvestris* after induction by insect oviposition. J. Chem. Ecol. 29:1235-1252.

Musser, RO, Hum-Musser, SM, Eichenseer, H, Pfeiffer, M, Ervin, G, Murphy, JB, Felton, GW (2002). Caterpillar saliva beats plant defences. Nature 116:599-600.

Popp, MP, Johnson, JD, Lesney, MS (1995). Characterization of the induced response of slash pine to inoculation with bark beetle vectored fungi. Tree Physiol. 15:619-623.

Raffa, KF (1991). Induced defensive reactions in conifer-bark beetle systems. In: Tallamy, DW, Raupp, MJ (eds), Phytochemical Induction by Herbivores. John Wiley & Sons, New York, pp. 245-276.

Rafii, ZA, Dodd, RS, Zavarin, E (1996). Genetic diversity in foliar terpenoids among natural populations of European black pine. Biochem. Syst. Ecol. 24:325-339.

Rezzi, S, Bighelli, A, Mouillot, D, Casanova, J (2001). Composition and chemical variability of the needle essential oil of *Pinus nigra* subsp. *laricio* from Corsica. Flavour Fragr. J. 16:379-383.

Roussis, V, Petrakis, PV, Ortiz, A, Mazomenos, BE (1995). Volatile constituents of needles of five *Pinus* species grown in Greece. Phytochemistry 39:357-361.

Sadof, CS, Grant, GG (1997). Monoterpene composition of *Pinus sylvestris* varieties and susceptible to *Dioryctria zimmermani*. J. Chem. Ecol. 23:1917-1927.

Schütt, P, Schuck, HJ, Lang, UM (1992). Handbuch und Atlas der Dendrologie. Ecomed, Landsberg/Lech.

Sjödin, K, Persson, M, Fäldt, J, Ekberg, I, Borg-Karlson, A-K (2000). Occurrence and correlations of monoterpene hydrocarbon enantiomers in *Pinus sylvestris* and *Picea abies*. J. Chem. Ecol. 26:1701-1720.

Sjödin, K, Persson, M, Borg-Karlson, A-K, Norin, T (1996). Enantiomeric compositions of monoterpene hydrocarbons in different tissues of four individuals of *Pinus sylvestris*. Phytochemistry 41:439-445.

Sjödin, K, Persson, M, Norin, T (1993). Enantiomeric compositions of monoterpene hydrocarbons in the wood of healthy and top-cut *Pinus sylvestris*. Phytochemistry 32:53-56.

Sjödin, K, Schroeder, LM, Eidmann, HH, Norin, T, Wold, S (1989). Attack rates of scolytids and composition of volatile wood constituents in healthy and mechanically weakened pine trees. Scand. J. For. Res. 4:379-391.

Sullivan, BT, Pettersson, EM, Seltmann, KC, Berisford, CW (2000). Attraction of the bark beetle parasitoid *Roptocerus xylophagorum* (Hymenoptera: Pteromalidae) to host-associated olfactory cues. Environ. Entomol. 29:1138-1151.

Sullivan, BT, Berisford, CW (2004). Semiochemical from fungal associates of bark beetles may mediate host location behavior of parasitoids. J. Chem. Ecol. 30:703-717.

Trapp, S, Croteau, R (2001). Defensive resin biosynthesis in conifers. Annu. Rev. Plant Physiol. Plant Mol. Biol. 52:589-724.

Trewhalla, KE, Leather, SR, Day, KR (1997). Insect induced resistance in lodgepole pine: effects on two pine feeding insects. J. Appl. Ent. 121:129-136.

Valterová, I, Sjödin, K, Norin, T (1995). Contents and enantiomeric compositions of monoterpene hydrocarbons in xylem oleoresins from four *Pinus* species growing in Cuba. Comparison of trees unattacked and attacked by *Dioryctria horneana*. Biochem. Syst. Ecol. 23:1-15.

van den Dool, J, Kratz, PD (1963). A generalization of the retention index system including linear programmed gas-liquid partition chromatography. J. Chromatogr. 11:463.

Wold, S, Albano, C, Dunn III, WJ, Esbensen, K, Geladi, P, Hellberg, S, Johansson, E, Lindberg, W, Sjøstrøm, M, Skagerberg, B, Wikström, C, Öhman, J (1989). Multivariate data analysis: Converting chemical data tables to plots. Intell. Instrum. and Comput.197-216.

Zivojinovic, D (1954). *Diprion pini* L. History of the gradation and consequences of the defoliation it caused on the Maljen. Plants Protection (Beograd) 24:3-19.

# Chapter 7

# Defence in Pines (*Pinus* spp.) to Biotic Stress by Herbivorous Insects

**Abstract.** Pines are subject to herbivory by a wide range of arthropods attacking many different tissues of the tree ranging from the roots over the trunk up to the needles and cones in the crown. To deter damage by herbivore, pines have evolved complex defence mechanisms which are reviewed here with special respect to herbivorous pine sawflies (Diprionidae, Hymenoptera) attacking needle tissue.

Physical defensive devices of pine will be briefly outlined. Toughness of pine tissue by lignification may provide an effective physical barrier against herbivores. Sticky resin exudates may just entangle and immobilize attacking herbivores arthropods. Desiccation of needles and their subsequent abscission in response to herbivory may be a costly, but effective strategy to remove the attackers from the tree. Also formation of necrotic tissue at the site of attack (hypersensitive response) may effectively isolate the attacker.

The major focus of this review is on chemical defence of pine by resin. This defensive strategy is considered from a biochemical, molecular and especially ecological point of view. The variability of resin constituents is discussed in context with its defensive effects. Constitutive defence in pine is contrasted with induced responses in pine and other plants. Rapid induced responses and long-term response to attackers of pine is considered. Finally, costs of defence in pine are addressed.

**Keywords.** oleoresin, monoterpenes, sesquiterpenes, diterpenes, induced defence, constitutive defence, rapid and long term responses, sawfly.

**Introduction**

Pines (*Pinus* spp.) are the largest genus of the conifer family Pinaceae consisting of more than 100 species and many varieties and subspecies (Langenheim 2003; Richardson and Rundel 1998). Data obtained from fossilized cones show that pines have evolved by the Lower Cretaceous (Richardson and Rundel 1998). Pines are mainly distributed throughout the temperate regions of the northern hemisphere and occupy a wide range of climatically different habitats (Langenheim 2003; Richardson and Rundel 1998). They especially prevail in regions with adverse conditions as well as in first succession stages (Keely and Zedler 1998).

Pines need to cope with a wide range of attackers. Some pine species (*P. banksiana*, *P. contorta*, *P. monticola*, *P. ponderosa*, *P. resinosa*, *P. strobus*, *P. sylvestris*) are attacked by more than 200 different insect species as well as various birds, mammals and fungi (de Groot and Turgeon 1998). Pines may defend against these attackers by both physical and chemical devices. Often it is difficult to separate these traits, since the physical devices need a chemical base, such as e.g. lignification for providing a physical barrier or sticky components of resin for entangling a herbivorous arthropod. Members of the Pinaceae synthesize copious amounts of viscous oleoresins consisting of terpenoid secondary metabolites (Gershenzon and Croteau 1991; Langenheim 1994) which provide a versatile set of defences against the numerous attackers. A plethora of studies has investigated the effects of resins and their variability on different herbivores and vice versa (e.g. Géri et al. 1993; Phillips and Croteau 1999; Raffa and Berryman 1987).

Furthermore, both physical and chemical defensive mechanisms in pine may be divided – like in other plants - into constitutively (preformed) ones that are permanently present and induced responses upon herbivore attack (Karban and Baldwin 1997; Larsson 2002; Trapp and Croteau 2001). While constitutive and inducible resistance mechanisms against herbivorous insects have been thoroughly described for other conifer genera like true firs (e.g. Phillips et al. 1999; Bohlmann and Croteau 1999), Douglas-fir (e.g. Clancy 2002), and spruces (e.g. Alfaro et al. 2002), a similar comparative study on pines is apparently lacking although these tree species has extensively been investigated. Here we give an overview of pine defence structures and mechanisms as well as their ecological roles in pines. These can be constitutively present and serve as a principal physical or chemical barrier against herbivores and pathogens. We will also summarize knowledge on how herbivory induces changes in the pine resin chemistry and conversely, how herbivores are affected by changes in the resin composition.

Numerous studies of chemically mediated defence in pine addressed the interactions of pines and scolytid beetles as well as their associated fungi. Very informative reviews of these interactions have been written by e.g. Lieutier (2002), Paine et al. (1997), Phillips and Croteau (1999), Raffa (1991, 2001), and Trapp and Croteau (2001). Here, we consider these interactions between pine, scolytids, and fungi only to exemplify special features and mechanisms regarding chemical defence in pines. Instead, this review focuses on physical and chemical defensive mechanisms of pine against conifer sawflies. Detailed reviews of conifer sawflies biology and chemical ecology have been given e.g. by Anderbrant (1999), Coppel and Benjamin (1965), Knerer and Atwood (1973), and Wagner and Raffa (1993), and will not be addressed here.

## 1. Physical Defence

### 1.1 "Tough Defence"

The toughness of pine tissue represents a first barrier for arriving herbivores that want to deposit eggs or want to feed. Tougher pine needles were found to be more resistant to attack by *Neodiprion fulviceps* sawflies and also negatively influenced the feeding success of the larvae (Wagner and Zhang 1993). On the other hand, the sawfly *D. pini* preferred to lay egg on tougher needles of Scots pine (*P. sylvestris*) (Pasquier-Barre et al. 2000). Mature pine trees possess a thick and lignified bark. The bark protects the vascular system against mechanical damage and serves as a constitutive physical barrier against non-specialized pathogens and herbivores

(Lieutier 2002; Schwerdtfeger 1981; Sitte et al. 1991). In particular, lignin has been shown to act as a chemical and physical barrier against herbivorous insects (Wainhouse et al. 1990).

### 1.2 "Sticky Defence"

Anatomical differences like a higher number of resin ducts are known to reduce the attack rate of the pine needle miner, *Exoteleia pinifoliella,* and of the white pine weevil (*Pissodes strobi*) (Hanover 1975). The anatomical structure of resin ducts differs significantly within the Pinaceae (Fahn 1979; Trapp and Croteau 2001, Wu and Hu 1997). Members of the genus *Pinus* have the most complex network of non-constricted resin ducts distributed in the cortex, primary and secondary xylem of stems and mesophyll (Trapp and Croteau 2001; Wu and Hu 1997). The primary body of *P. halepensis* consists of three independent types of resin ducts: (1) the root-hypocotyl system, (2) the shoot system, and (3) the needle system (Werker and Fahn 1969). The secondary ducts constitute one system in the secondary xylem and phloem. In the xylem there are two-dimensional interconnected vertical and horizontal resin ducts, whereas only horizontal ducts occur in the phloem (Werker and Fahn 1969). Connections between primary and secondary resin ducts have only been found between vertical ducts and secondary radial ducts. This might be a reason for different resin compositions between organs (see below; Langenheim 2003; Werker and Fahn 1969). Only the epithelial cells of the secondary xylem resin ducts of pines remain alive and thin-walled (Wu and Hu 1997). Due to the highly developed net of resin ducts, pines produce and accumulate large amounts of oleoresin constitutively (Lewinsohn et al. 1991a, Trapp and Croteau 2001).

In response to mechanical wounding, anatomical modifications are induced in conifers. The number of resin ducts may also be increased in response to attack. The production of so-called traumatic resin ducts (TRD) can be elicited by many stimuli, including mechanical wounding, pathogens (e.g. Cheniclet 1987) or exogenously applied elicitors such as chitosan (Lieutier 2002). Traumatic resin ducts are clearly distinguishable from constitutive resin canals and contain resin that has higher monoterpene concentrations compared to constitutive resin (Alfaro et al. 2002).

Though the induction of traumatic resin ducts seem to be a generalized defence system in conifers, the most extensive studies have been conducted with spruces (*Picea* spp.) as model plants (Alfaro et al. 2002; Franceschi et al. 2000; McKay et al. 2003; Nagy et al. 2000; Tomlin et al. 1998). In contrast, in response to wounding pines form additional resin ducts which are anatomically indistinguishable from constitutive resin ducts and sometimes are improperly referred as TRD (Cheniclet 1987; Fahn and Zamski 1970).

Upon exposure to the atmosphere, mono- and sesquiterpenes of the resin exsudating from pine wounds evaporate, leaving a semicrystalline mass of resin acids that oxidatively polymerise to form a sticky and later more and more hardened barrier that seals the wound, often trapping insect invaders and microbial pathogens in the matrix (Phillips and Croteau 1999).

**1.3 "Dry Defence"**

In *Pinus resinosa*, egg deposition by *Neodiprion lecontei* (Hymenoptera, Diprionidae) induces desiccation of needles carrying the sawfly eggs (Codella and Raffa 2002). Sawfly eggs need the humidity within the pine needles, and thus, pine needle desiccation was found to cause high egg mortality (Codella and Raffa 2002). It has been controversially discussed whether premature leaf desiccation or abscission is an active defence response by the plant or is the result of mechanical damage of the plant tissue (Williams and Whitham 1986; Stiling and Simberloff 1989). However, according to Björkman and Larsson (1991) abscission of needles of *P. sylvestris* bearing egg of *N. sertifer* or whole egg batches has never been observed.

While abscission of needles requires formation of necrotic tissue at the needle base, pine cells may also die and desiccate locally very restricted to the site of attack. For example, after the colonization of bark beetles or inoculation with beetle-associated fungi, pines show hypersensitive responses (HR) including cell necrosis and initiation of new impermeable cell layers at the site of attack (Lieutier 2002, Paine et al. 1997, Raffa 1991). HR is induced by the mechanical damage caused by the tunnelling activity of the beetles and is amplified by the structural and metabolic properties, of associated fungi, e.g. elicitors like chitosan, a fragment of the fungal cell wall (recent reviews by Lieutier 2002, Paine et al. 1997). However, bark beetles induced HR are always only part of multiple response mechanisms of pines which are especially accompanied with chemical defence reactions.

## 2. Chemical Defence

One of the most important defence mechanisms of pines is the production of oleoresin (resin, pitch). Oleoresin is produced in large amounts by special secretory tissues and is stored in a network of specialized resin ducts, which are located throughout the wood, bark, roots and needles (Fahn 1979; Gijzen et al. 1993; Trapp and Croteau 2001). To understand the mode of action of wounding induced chemical responses in pine, detailed studies of the biosynthesis of the oleoresin constituents is necessary, both from a chemical and molecular perspective.

### 2.1 Biosynthesis of Resin

When considering the biochemistry of oleoresin terpenes, isopentenyl diphosphate (IPP), traditionally called isopentenyl pyrophosphate, forms the central intermediate. IPP can be formed via two different pathways, which operate in two different plant compartments (Bernard-Dagan 1982):

(a) In the classic acetate/mevalonate pathway (mevalonic acid pathway) IPP is formed by three molecules of acetyl CoA with mevalonate as intermediate (Chappell 1995). This pathway operates in the cytosol compartment in which sesquiterpenes and triterpenes are formed (Gleizes et al. 1980; Trapp and Croteau 2001).

(b) The alternate mevalonate-independent pathway (also referred as DOXP or DXP pathway, pyruvate-gylceraldehyde-3-phosphate pathway, or MEP pathway) is initiated by pyruvate and glyceraldehyde 3-phosphate. These components form 1-deoxy-D-xylulose-5-phosphate (DOXP) that oxidizes to 2-C-methyl-D-erythritol-4-phosphate (MEP). The latter is transformed to IPP. This pathway operates in the plastids to provide monoterpenes, diterpenes, and tetraterpenes (Lichtenthaler 1999, 2000; Lichtenthaler et al. 1997).

IPP and its isomer, dimethylallyl diphosphate (DMAPP), are the actual five-carbon building precursors for the formation of larger terpenoid molecules (Langenheim 2003). DMAPP is condensed with IPP molecules which is catalysed by specific prenyltransferases. Depending on how many IPP units are combined with DMAPP different types of terpenoids are synthesized. If one IPP is added with DMAPP geranyl diphosphate (GPP) the precursor of the monoterpenes ($C_{10}$) is formed. Addition of one or two further IPP units leads to farnesyl diphosphate (FPP), which is the precursor of sesquiterpenes ($C_{15}$) and diterpenes ($C_{20}$), geranylgeranyl diphosphate (GGPP), respectively (Phillips and Croteau 1999, Trapp and Croteau 2001).

Numerous molecular studies of terpenoid biosynthesis focus on terpenoid synthases, often also called cyclases because many reaction products are cyclic. These enzymes convert the respective acyclic precursors to the myriad of monoterpenes, sesquiterpenes, and diterpenes. Up to now, more than 30 terpene synthase encoding genes have been isolated (Langenheim 2003). Phylogenetic analyses based on the amino acid sequences revealed that conifer terpene synthases are more closely related to each other than to their angiosperm counterparts (Bohlmann et al. 1998). Moreover, pine synthases producing similar terpenes are more closely related to each other within the same species than to the same synthases of grand fir (Savage et al. 1994). Many terpene synthases are known to produce a single product, but there are several enzymes that produce multiple products (Croteau 1987; Savage et al. 1994).

The majority of biochemical studies of terpene synthases has been conducted with grand fir (*Abies grandis*) as model plant, especially because this conifer shows a strong wound-induced resinosis, thereby offering significant advantage for biochemical studies (Bohlmann et al. 1998, Bohlmann and Croteau, 1999). Only few terpene synthases from pines have been investigated in detail yet. Pines have the highest monoterpene synthase activity compared to other conifer species, which corresponds well with the most highly developed resin duct system and with the highest constitutive levels of monoterpene concentrations (Lewinsohn et al. 1991b; Litvak and Monson 1998). Four synthases isolated from xylem of *P. contorta* formed multiple monoterpene products as sabinene, β-phellandrene, 3-carene and β-pinene (Savage et al. 1994, 1995). When considering sesquiterpene synthases, an (*E*)-β-farnesene synthase that has been purified from maritime pine (*Pinus pinaster*) is the only sesquiterpene synthase characterized from a pine species so far (Salin et al. 1995). The enantiomers of a terpenoid component are known to be synthesized by either the same enzyme, as for α-pinene in *P. contorta* (Savage et al. 1995), or by different enzymes as in *P. taeda* (Phillips et al. 1999, 2003). The enantiomeric forms of a terpenoid compound may have very different biological activities (e.g. Byers 1995; Erbilgin and Raffa 2001a; Sjödin et al. 2000; Wibe et al. 1998).

### 2.2 Pine Resin Composition

Pine resin is composed of a complex blend of terpenoids consisting of a volatile turpentine fraction (monoterpenes, sesquiterpenes) and a less volatile rosin fraction (diterpene resin acids and phenolic compounds), and their derivatives (Fahn 1979; Gershenzon and Croteau 1991; Langenheim, 1994; Mirov 1967). The volatile constituents are toxic and/or repellent to many herbivores. They may also serve as solvent for the mobilization of the resin acids (reviewed by Langenheim 1994, Trapp and Croteau 2001).

Monoterpenes commonly occur as hydrocarbons with few oxidized forms. Most of them possess a high volatility. Sesquiterpenes represent the largest terpenoid class and possess similar effects on herbivores as described for monoterpenes (see below) though they are generally produced to a far lesser extend. Pine resin contains also diterpene acids that are produced in similar amounts as mono- and sesquiterpenes taken together (Trapp and Croteau 2001). Diterpene acids are of very low volatility. They are also toxic or deterrent for some herbivore insects (Géri et al. 1993).

### 2.3 Variability of Pine Resin

Numerous studies compare qualitatively and quantitatively the pattern of terpenoid components and their enantiomers (e.g. Fäldt et al. 2001; Sjödin et al. 1996, 2000). The composition and amounts of oleoresin varies significantly depending on numerous factors. In numerous studies investigating resin contents and headspace volatiles of different pine species, α-pinene and β-pinene (Fig. 1) were always among the five major monoterpenes (e.g. Macchioni et al. 2003; Petrakis et al. 2001; Roussis et al. 1995). But also myrcene, 3-carene, limonene, and β-phellandrene were often detected in considerable amounts (Fig. 1). Quantities of the monoterpene 3-carene are known to extremely vary in *P. sylvestris*. Therefore, trees can be grouped in either high (> 10 % of total terpenes) or low (< 10 % of total terpenes) 3-carene containing trees (Hiltunen et al. 1975; Manninen et al. 2002; Sjödin et al. 1996, 2000). The composition of sesquiterpenes has not been investigated to the same extent as for monoterpenes. In most of the studies, germacrene D, β-caryophyllene and δ-cadinene were found to be the dominating sesquiterpenes in pine tissue (Fig. 1). Common diterpenoid resin acids of pines are pimaric, sandaracopimaric, isopimaric, palustric + levopimaric, dehydroabietic, abietic, and neoabietic acids (Fig. 1) (Coppen et al. 1998, Manninen et al. 2002, Nerg et al. 2004, Sallas et al 2001).

The composition and amounts of oleoresin varies significantly between pine species (Fäldt et al. 2001; Macchioni et al. 2003; Petrakis et al. 2001; Roussis et al. 1995, Tiberi et al. 1999). Terpene composition displays also considerable variation among populations in which monoterpenes showed generally greater population differentiations than diterpenoids (Coppen et al. 1998; Forrest et al. 2000; Latta et al. 2003; Manninen et al. 2002; Rafii et al. 1996). While particular pine species contain a characteristic resin composition, the terpenoid fractions may vary considerably between individuals (Sjödin et al. 1996), and furthermore between different tissues within single pine trees (Hanover 1975, Latta et al. 2000; Manninen et al. 2002; Roberts 1970, Sjödin et al. 1996). In *P. sylvestris*, Sjödin et al. (1996) divided tissue samples into (1) trunk and root xylem and root phloem, (2) needles, and (3) branch tissues (except needles) based on similarities of the relative amounts of monoterpenes and their enantiomeric composition. Manninen et al. (2002) showed that in *P. sylvestris* the total monoterpene concentration was five-fold higher in needles than in wood tissue, whereas this was not the case in *P. ponderosa* (Latta et al. 2000). However, in ponderosa pine only the terpenes of needle tissues were significantly affected by the crown position (north/south side) which were weakly correlated with the concentrations of other tissues (Latta et al. 2000). Moreover, the monoterpene concentrations and synthase activity differ significantly between the top and lower half of pine needles (Litvak and Monson 1998). Overall, the terpenoid pattern of pine needles is the most variable tissue within a pine tree and usually differs from the one of other pine tissue.

In addition, the great variability in the concentration and composition of needle resin likely reflects the quantities and composition of volatile terpenoids in the headspace of a pine tree, because the needles are supposed to emit the majority of terpenoids from their great surface. Moreover, the emission of pine volatiles varies strongly dependent on several abiotic factors as temperature, humidity and partly photosynthetic active radiation (Komenda, 2001; Lerdau and Gray 2003; Shao et al. 2001; Tingey et al. 1991). Thus, there is a high qualitative and quantitative chemical variability of the volatile pattern within a pine tree. Though under genetic control, resin concentration and composition is also influenced by other physiological factors, such as e.g. water availability, nutrient stage, developmental stage (e.g. Lombardero et al. 2000, Turtola et al. 2003), and environmental factors (Langenheim 1994; Nerg et al. 1994). Hence, the intra- and interplant chemical variation is spatially and temporally heterogeneous, providing pines with a defensive mosaic effect, which is assumed to be the best defensive strategy because pines are attacked by several pests on many different spatial and temporal scales (Langenheim 2003).

## 2.4 Ecological Roles of Chemical Defence

In this section we give an overview of how resistance in pines against herbivores is obtained by constitutively produced and induced defensive chemicals. Pine resins can have direct detrimental effects on herbivores or they can also have an indirect effect by involving e.g. natural enemies. We will focus on chemically mediated interactions between pines and pine sawflies (Hymenoptera, Diprionidae), but refer also to other plant-herbivore systems where appropriate.

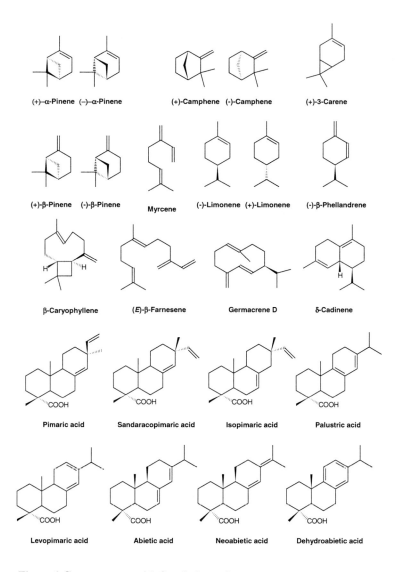

**Figure 1** Common terpenoids found pine resin

**2.4.1 Effects of constitutive (preformed) resin defence**

Pines defend themselves against herbivores by producing large amounts of resin constitutively. Upon wounding, resin is translocated to the injured site and invaders like stem boring insects and their associated fungi are essentially flushed (pitchted) out by the flow of resin (Trapp and Croteau 2001). In vigorous trees, both flow capacity and resin exudation pressure is very high, providing an effective physical barrier. Furthermore, toxic resin acids that exudes from the wound forms a semicrystalline plug when the volatile compounds evaporate.

Thus, for non-specialised insects pine resin might function qualitatively by either deterring or killing the animal (Feeny 1992). Regarding more adapted insects, the defensive effect of resin is strongly dependent on the concentration and composition of resin constituents. Several studies showed that the performance of specialised pine sawflies (Hymenoptera, Diprionidae), which are severe defoliators on pines, is significantly reduced on particular pine species or genotypes of the same species from different provenances (Auger et al.1994a; Barre et al. 2002, 2003; Olofsson 1989; Pasquier-Barre et al. 2000, 2001; Trewhalla et al. 2000). Though not all such negative effects of a particular pine to the sawflies can be explained by resin components, several cases showed a correlation between pine terpenoid content and host tree suitability. For example, in the European pine sawfly *Neodiprion sertifer* survival and performance of larvae as well as the female cocoon weight and fecundity are negatively correlated with concentration of resin acid (Björkman 1997; Larsson et al. 1986, 2000). On the other hand, larval performance and development of sawflies show a poor correlation to the monoterpene composition though some compounds might be active (Codella and Raffa 1995a, Géri et al. 1993, Lyytikäinen 1994).

Interestingly, the oviposition behaviour of pine sawflies is only weakly correlated with the larval performance. Female sawflies do not necessarily prefer to lay eggs on trees with low resin acid or taxifolin contents which would favour the performance of larvae (Björkman et al. 1997; Pasquier-Barre et al. 2000; Saikkonen et al. 1995). This behaviour might be maladaptive at a first glance but this way the offspring might escape high mortality due to natural enemies because predation is significantly lower when larvae feed on pine foliage containing high levels of resin acids (Björkman and Larsson 1991). In some sawfly species, the monoterpene composition of pine foliage seems not to significantly influence the oviposition behaviour (Codella and Raffa 1995b; Pasquier-Barre et al. 2000). However, *Diprion pini* females were deterred from oviposition by regurgitate of conspecific larvae containing a quantitatively different monoterpene pattern than pine resin which did not act as oviposition deterrent (Hilker and Weitzel 1991, Weitzel 1991).

Many insects have evolved behavioural or physiological adaptations to cope with the plant's chemical defence mechanisms and use them for own purposes thereby attenuating the resistance of their host trees (Litvak and Monson 1998; McCullough and Wagner 1993; Raffa 1991). For example, larvae of conifer sawflies possess a pair of foregut diverticulae where pine-derived terpenoids are sequestered (Eisner et al. 1974, Björkman and Larsson 1991). The content of the diverticulae is regurgitated upon disturbance which has been shown to repel different predators (Codella and Raffa, 1993), but also conspecific females (Hilker and Weitzel 1991). The regurgitated amount and efficacy of the fluid in repelling ants is significantly affected by the concentration of resin acids from the host plants (Codella and Raffa 1995a). In contrast, sawfly larvae that fed upon pine foliage with high resin acid concentration were not better protected against parasitism and predation in the cocoon stage (Björkman and Gref 1993). Thus, producing constitutively high amounts of resin acids shows a trade-off for pines between being well-defended directly against conifer sawflies but reducing the efficacy of indirect defence mechanisms because of the detrimental effects on carnivores.

There is ample evidence that volatile terpenes play an essential role in direct and indirect defence mechanisms of pines against herbivorous insects other than sawflies (Hanover 1975, Langenheim 1994). Much work has focused on the attractive activity of volatiles monoterpenes for many insects rather than on their repellent properties. For example, pine monoterpenes appear to be important cues for host finding and oviposition in pyralid moths of the genus *Dioryctria* (Kleinhentz et al. 1999; Sadof and Grant 1997; Shu et al. 1997; Valterová et al. 1995). Especially the proportion of some monoterpenes were different between resistant and susceptible host trees of *Dioryctria* moths. Pines that are hardly attacked by a particular herbivore might escape herbivory by just releasing a volatile blend non-attractive to the herbivore.

Monoterpenes have complex and dose-dependent functions in conifer-insect interactions (e.g. Miller and Borden 1990, 2000; Nordlander 1990). High concentrations of monoterpenes can have detrimental or even toxic effects on bark beetles and their fungi (Raffa and Smalley 1995), but many bark beetles are also known to use monoterpenes from the host plant as kairomones either alone or in combination with aggregation pheromones (Byers 1995, Miller and Borden 1990). Furthermore, bark beetles have the ability to transform many monoterpenes to oxygenated derivatives, which then function as sex or aggregation pheromones, thereby detoxifying and undermining the host defence (Byers 1995, Phillips and Croteau 1999). For example, in the pine feeding *Ips pini*, α-pinene was not attractive *per se* but, depending on the concentration, was able to synergise or inhibit the attractiveness of the beetles aggregation pheromones, respectively (Erbilgin and Raffa 2000, Erbilgin et al. 2003). Thus, a high ratio of α-pinene to bark beetle pheromone might be an important component of integrated host defences against bark beetles (Erbilgin et al. 2003). Likewise, α-pinene and 3-carene were not attractive themselves, but synergise the attractiveness of the *I. pini* aggregation pheromones to its predominant predator, *Thanasimus dubius* F. (Coleoptera: Cleridae) (Erbilgin and Raffa 2001a, Erbilgin et al. 2003).

It is well known that carnivorous arthropods use pheromones of their pine feeding hosts for host location (e.g. Erbilgin and Raffa, 2001b; Hilker et al. 2000; Miller and Borden 2000; Raffa and Dahlsten 1995). It might be adaptive for a carnivorous arthropod to respond to a combination of host pheromones and pine volatiles when foraging for prey. Pheromones are highly reliable regarding the presence of a suitable host, but relatively difficult to detect due to minute amounts. Plant terpenes are emitted in large amounts which are easy to detect. However, they show a high genotypic and phenotypic variation reducing the reliability for a foraging predator or parasitoid (Vet and Dicke 1992). The combinatory use of pheromones and constitutive plant volatiles might be a suitable strategy for a carnivore to locate the prey by reliable and detectable cues.

### 2.4.2 Effects of induced resin defence

A further strategy for carnivores to detect pine herbivores is to rely on herbivore-induced pine volatiles. Several studies show that volatiles from herbivore-infested pines were significantly more attractive for parasitoids than uninfested ones (Hilker et al. 2002; Sullivan and Berisford 2004; Sullivan et al. 2000; Völkl 2000). Pine volatiles were still attractive when host and their products were removed (Sullivan and Berisford 2004; Sullivan et al. 2000). These studies indicate that pine and pine resin have changed their volatile pattern in response to herbivory so that carnivores were attracted.

Recently, it was shown that the egg deposition of the pine sawfly *Diprion pini* induces changes in the volatile pattern of *P. sylvestris* twigs that are attractive to females of *Chrysonotomyia ruforum* Krausse (Hymenoptera, Eulophidae), an egg parasitoid of *D. pini* (Hilker et al. 2002). The induction of volatiles attractive for the egg parasitoids was systemic, i.e. parts of the same twig that carried no eggs were also attractive to *C. ruforum*. Volatiles of pine twigs that were artificially wounded to mimic the mechanical damage the female sawfly inflicts with her ovipositor prior to egg deposition did not attract the egg parasitoids. Thus, pines "recognise" the attacking herbivore and respond to this in a different manner, which confirms other studies showing that wounding did not elicit the same responses in plants as herbivores or pathogens (see next paragraph below). Analysis of the volatile pattern of pine induced by egg deposition and artificially wounded control twigs revealed no qualitative but quantitative alterations (Mumm et al. 2003). No significant changes in the monoterpene composition were detected, but the sesquiterpene $(E)$-$\beta$-farnesene was released in significant higher amounts by the oviposition-induced pine twigs compared to the respective controls (Mumm et al. 2003). Further experiments showed that $(E)$-$\beta$-farnesene was able to attract female egg parasitoids when offered in combination with non-induced pine volatiles (Chapter 4). Thus, just a very minute, but significant quantitative change of the ratio of a single monoterpene within the blend of numerous other terpenoid pine components results in the attraction of an egg parasitoid which seems to be highly adapted to a very finely tuned change of volatiles. The plant's defensive response to insect egg deposition can be interpreted as a preventive defence mechanism since they respond to the herbivore attack prior being damaged by the feeding sawfly larvae (Hilker and Meiners 2002).

Also other studies show that pines respond to herbivory or pathogen attack only by minor changes of the terpenoid pattern. For example, Raffa and Smalley (1995) found only small differences between the monoterpene composition of induced and constitutive monoterpenes in *P. resinosa* and *P. banksiana* after fungal inoculation. The induced responses in pines are weak compared to other conifer genera with less developed preformed resin production and accumulation. Extracts from stem tissue of several pine species showed a high constitutive monoterpene synthase activity which were not inducible by wounding within a time period of 1-10 days (Lewinsohn et al. 1991a,b; Phillips et al. 1999). Several studies showed that herbivory or wounding did not cause consistent changes in the terpene pattern in pines (Fäldt et al. 2001; Sjödin et al. 1989; Valterová et al. 1995). However, in *P. caribaea* increased relative amounts of (-)-α-pinene and of both enantiomeres of β-pinene were found in trees attacked by *Dioryctria horneana* compared to uninfested ones (Valterová et al. 1995).

However, like in other plant - herbivore interactions, herbivores were also found to adapt to induced responses of pine. Defoliation of *P. resinosa* either artificially or by natural feeding of *N. sertifer* larvae improved the quality of foliage for conspecific larvae a few days later (Krause and Raffa 1995). This induced suitability of previously defoliated pine seedling was, however, dependent on the larval age and the defoliation intensity. Litvak and Monson (1998) showed a significant increase in monoterpene cyclase activity in needles of *P. contorta* and *P. ponderosa* after real and simulated herbivory. The authors further demonstrated that wounding induced only an increase in the α-pinene level of needles, whereas the total needle monoterpene concentration decreased due to a higher volatilisation of monoterpenes.

Thus, the induced production of monoterpenes serves probably to replace the constitutive pool that is lost to volatilisation (Litvak and Monson 1998).

Systemically induced defence responses as demonstrated in *P. sylvestris* after egg deposition of sawflies results in stimulation of defence mechanisms in uninfested structures spatially away from the infestation site. This may concern parts only a few centimeters away, structures within the same modular unit or the whole tree. The latter case might be adaptive for resistance traits acting directly against herbivores, thus protecting the whole tree after an initial attack (Auger et al. 1994b; Heidger and Lieutier 2002). Conversely, as an indirect mechanism by attracting natural enemies it may be a long-range cue at best, whereas it would be less suitable to guide carnivores to the hosts within a tree. Recently, a study by Wallin and Raffa (1999) showed that defoliation by the jack pine budworm *Choristoneura pinus pinus* induced alterations in the terpene compositions of phloem in *P. banksiana,* indicating that systemic responses are not restricted to particular tissues.

The induced effects we described here can be considered so called rapid (short-term) induced responses which directly effect the attacking herbivores or at least the same generation as reported. Many studies investigated whether plants also show delayed induced responses, i.e. if plants are more resistant to future herbivore generations after previously been attacked (Haukioja and Honkanen 1997). Delayed induced resistance has been well documented in mountain birch *Betula pubescens* against the geometrid moth *Epirrita autumnata* (Haukioja and Honkanen 1997). However, other studies showed miscellaneous delayed effects of induced responses. For example, the performance of conifer sawflies was not reduced when they fed on foliage from *P. sylvestris* trees that had been previously defoliated (Niemelä et al. 1984, 1991). The needles of previously defoliated trees tended to be even more suitable to sawfly larvae. An induced suitability to sawflies after previous defoliation was generally confirmed by Raffa et al. (1998), but the results were not consistent regarding the intensity of defoliation, insect sex and measured performance parameter as well as tree age. In comparison to deciduous trees, conifers store carbon and nutrients in the needles, so that defoliation results in a change of the carbon : nutrient balance and in a reduction of the carbohydrate pool used for carbon based chemical defence (Bryant et al. 1988; Raffa et al. 1998; Smits and Larsson 1999). The increased suitability of pine after sawfly defoliation may explain why outbreaks of conifer sawflies can last several seasons in the same area and are controlled rather by carnivores and pathogens which can cause considerable mortality (Larsson et al. 1993; Niemelä et al. 1984). Thus, though there might be long term inducible responses in pines against herbivory there is little support that these responses result in an increased resistance. This general view is confirmed by a meta-analysis by Nykänen and Koricheva (2004) who found a comparably weaker induced resistance in Scots pine than in other trees. However, Larsson (2002) points out that the performance of usually less investigated life stages, e.g. the herbivore's eggs, or the adult (e.g. the oviposition behaviour) might be affected by induced responses.

## 2.5 Specificity of Induced Responses in Pines – the Role of Potential Elicitors

Herbivores generally injure the conifer host often in a very characteristic way when they feed or lay eggs on it. Therefore, when investigating inducible responses, plants are wounded either artificially or naturally by the herbivores and then the effects are compared with control plants. Artificially wounding normally tries to simulate the mechanical damage inflicted by a herbivore in a suitable way, e.g. by clipping needle foliage with scissors to mimic the feeding of sawfly larvae (e.g. Litvak and Monson 1998; Lyytikäinen 1992) or by drilling stem tissue to simulate invading and tunnelling bark-beetles (e.g. Lewinsohn et al. 1991a). In several studies the measured effect after artificially wounding was similar to the effect of a natural attack by herbivores (e.g. Lyytikäinen et al. 1992). This indicates an induced plant response to mechanical tissue damage and not a specific response to the attacking herbivore. In fact, other studies showed that artificially wounding failed to adequately mimic the effects induced by herbivores which might be due to an unsuitable method of wounding (Baldwin 1990; Kulman 1971; Litvak and Monson 1998; Hilker et al. 2002).

On the other hand, plants can "recognise" intruders which is assumed to be mediated through eliciting signal molecules from the attacker that are transferred into the wounded plant tissue (e.g. Cervone et al. 1997; Ebel and Mithöfer 1998; Gatehouse 2002). For example, pines respond to attacking bark beetles with a hypersensitive response which also occurs in response to artificially wounding. However, the response to artificial wounding is weaker and different than to a real beetle attack (Lieutier 2002; Paine et al. 1997). The elicitors probably do not originate from the beetles themselves, but primarily from the fungi which are specifically associated with the bark beetles (Paine et al. 1997). The underlying mechanism is not well understood but experiments treating pine tissue with chitosan (a fragment from the fungal cell wall) showed an induce responses on a chemical and molecular level (Mason and Davis 1997; Miller et al. 1986). Although induced responses have shown some specificity according to the pathogenicity of the fungi, others seem to be rather non-specific (Lieutier 2002; Paine et al. 1997, Popp et al. 1995).

The elicitor responsible for the oviposition-induced emission of pine volatiles that are attractive for egg parasitoids was located in the oviduct secretion of the female sawflies (Hilker et al. 2002). The oviduct secretion which coats the eggs is transferred into the wounded plant tissue during egg deposition. Wounding *per se* did not induce the production of attractive volatiles but when oviduct secretion was applied into artificially wounded pine needles the production of volatiles attractive to the egg parasitoids was induced (Hilker et al. 2002, see Chapter 2). Further studies indicated that the elicitor within the oviduct secretion has proteinacous properties since the activity vanished after the oviduct secretion was kept frozen or treated with proteinase K (Chapter 2).

In contrast to elicitors isolated from plant pathogens, little is known about the mode of action of herbivore elicitors (Dicke and van Poecke 2002). The elicitor within the oviduct secretion of *D. pini* was only active in disrupted needle tissue but inactive when oviduct secretion was applied on the intact needle surface. Similar results were found in other systems where plant damage was necessary to maintain the activity of elicitors (Mattiacci et al. 1995; Meiners and Hilker 2000; Turlings et al. 1990).

## 3. Costs of Defence

Pines are involved in a myriad of mutualistic and antagonistic interactions with herbivores, carnivores, pathogens, and other plants. In this complex ecological context, many of these interactions likely influence and are influenced by terpenoid compounds from the resin. Thus, the selection pressure on pines to alter resin chemistry is multidimensional, and involves considerations of attraction or repulsion, toxicity, pheromone signalling and tritrophic interactions (Phillips & Croteau 1999). Pines therefore commonly possess multiple, finely tuned responses to attack by herbivores and pathogens.

The investment into induced responses needs to cover metabolic costs (e.g. Gershenzon 1994). Furthermore, Cipollini et al. (2003) claimed that also the costs of simply being ready to become induced, i.e. the costs of inducibility, need consideration. Moreover, possible ecological costs need to be taken into account, i.e. the effects of induced responses need to balance the effects on both the attacker and the third trophic level, the carnivores. In fact, direct and indirect defences may counteract, as some plant toxins are detrimental for the herbivore, but also affect natural enemies negatively (reviewed by Bottrell et al. 1998; Hare 2002). It has been predicted that constitutive and induced defences will be negatively correlated (e.g. Karban and Myers 1989) which was confirmed in a meta-analysis by Koricheva (2004).

Pines might "afford" a low inducibility because they produce copious amounts of oleoresin constitutively (Lewinsohn et al. 1991a). In response to herbivore damage or wounding constitutive ("primary") resin is translocated to the sites of attack (Gijzen et al. 1993) which is followed by an induced *de novo* synthesis and accumulation of terpenes ("secondary resin") around the infestation site (Cheniclet 1987; Raffa 1991).

This response is often accompanied by forming a necroting zone around the wounding site and a reactivation of pre-existing or production of new resin ducts (Cheniclet 1987; Raffa 1991).

Though pines produce and accumulate terpenoids constitutively in high amounts, several studies showed also showed that pines may change their terpenoid quantities and qualities in response to herbivory (see above). Koricheva (2004) argued constitutive and induced defence strategies are not functionally redundant and provide different benefits for the plant. We agree with this and further suggest that induced responses in pines may be especially important in tritrophic interactions as has been demonstrated in numerous other systems (reviewed by Dicke et al. 2003; Turlings et al. 2002).

Herbivore-induced volatiles can be very specific for the plant and herbivore species, though they are subject to a high phenotypic variability due to several physiological and environmental factors (Dicke 1999; Turlings et al. 1998, 2002; van den Boom et al. 2004). Many carnivores can adjust their behaviour to varying quantities and qualities of plant cues by learning these cues during an host encounter (Turlings et al. 1993; Vet et al. 1995). It has recently been shown that parasitoids might be a important selective force in the evolution of herbivorous forest insects that is strongly influenced by the host plant (Lill et al. 2002).

Many studies concerning herbivore induced defence responses in pines focused on alterations of monoterpenes (e.g. Litvak and Monson 1998; Raffa and Smalley 1995; Wallin and Raffa 1999). However, in many plants also sesquiterpenes are induced after herbivore feeding, egg deposition or elicitor treatment (e.g. Gouinguené et al 2003; Martin et al. 2003; Mumm et al. 2003; Röse and Tumlinson 2004; Turlings et al. 1991). Though they are emitted in lower amounts and are less volatile than monoterpenes, sesquiterpenes might be important olfactory cues for carnivores (Chapter 4). Both herbivores and parasitoids are known to be more sensitive to less volatile compounds, e.g. β-caryophyllene than to more volatile compounds as α-pinene (Park et al. 2001; Zhang et al. 2003). We suggest that future studies investigating induced responses in conifers should also take a close look on differences in the sesquiterpene composition.

## Acknowledgements

This study was supported by the Deutsche Forschungsgemeinschaft (DFG Hi 416/11-1,2).

## References

Alfaro, RI, Borden, JH, King, JN, Tomlin, ES, McIntosh, RL, Bohlmann, J (2002). Mechanisms of resistance in conifers against shoot infesting insects. In: Wagner, M, Clancy, K, Lieutier, F, Paine, T (eds), Mechanisms and Deployment of Resistance in Trees to Insects. Kluwer Academic Publishers, Dordrecht, pp. 105-130.

Anderbrant, O (1999). Sawflies and seed wasps. In: Hardie, J, Minks, AK (eds), Pheromones of Non-Lepidopteran Insects Associated with Agricultural Plants. CABI Publishing, Wallingfort, New York, pp. 199-226.

Auger, M-A, Géri, C, Allais, J-P (1994a). Effect of the foliage of different pine species on the development and on the oviposition of the pine sawfly *Diprion pini* L. (Hym., Diprionidae); II. Influence on egg laying and interspecific variability of some active secondary compounds. J. Appl. Ent. 117:165-181.

Auger, M-A, Jay-Allemand, C, Bastien, C, Géri, C (1994b). Quantitative variations of taxifolin and it's glucoside in *Pinus sylvestris* needles consumed by *Diprion pini* larvae. Ann. Sci. For. 51:135-146.

Baldwin, IT (1990). Herbivory simulations in ecological research. Trends Ecol. Evol. 5:91-93.

Barre, F, Goussard, F, Géri, C (2003). Variation in the suitability of *Pinus sylvestris* to feeding by two defoliators, *Diprion pini* (Hym., Diprionidae) and *Graellsia isabellae galliaegloria* (Lep., Attacidae). J. Appl. Ent. 127:249-257.

Barre, F, Milsant, F, Palasse, C, Prigent, V, Goussard, F, Géri, C (2002). Preference and performance of the sawfly *Diprion pini* on host and non-host plants of the genus *Pinus*. Entomol. Exp. Appl. 102:229-237.

Bernard-Dagan, C, Pauly, G, Marpeau, A, Gleizes, M, Carde, J-P, Baradat, P (1982). Control and compartmentation of terpene biosynthesis leaves of *Pinus pinaster*. Physiol. Veg. 20:775-795.

Björkman, C (1997). A dome-shaped relationship between host plant allelochemical concentration and insect size. Biochem. Syst. Ecol. 25:521-526.

Björkman, C, Gref, R (1993). Survival of pine sawflies in cocoon stage in relation to resin acid content of larval food. J. Chem. Ecol. 19:2881-2890.

Björkman, C, Larsson, S (1991). Pine sawfly defence and variation in host plant resin acids: a trade-off with growth. Ecol. Entomol. 16:283-290.

Björkman, C, Larsson, S, Bommarco, R (1997). Oviposition preferences in pine sawflies: a trade-off between larval growth and defence against natural enemies. Oikos 79:45-52.

Bohlmann, J, Meyer-Gauen, G, Croteau, R (1998). Plant terpenoid synthases: Molecular biology and phylogenetic analysis. Proc. Natl. Acad. Sci. 95:4126-4133.

Bohlmann, J, Croteau, R (1999). Diversity and variability of terpenoid defences in conifers: molecular genetics, biochemistry, and evolution of the terpene synthase gene family in grand fir (*Abies grandis*). In: Chadwick DJ, Goode JA (eds), Insect-Plant Interactions and Induced Plant Defence. Wiley & Sons, Chichester, pp. 132-145.

Bottrell, DG, Barbosa, P, Gould, F (1998). Manipulating natural enemies by plant variety selection and modification: a realistic strategy? Annu. Rev. Entomol. 43:347-367.

Bryant, JP, Tuomi, J, Niemelä, P (1988). Environmental constraint of constitutive and long-term, defenses in woody plants. In: Spencer, K (ed), Chemical Mediation of Coevolution. Academic Press, San Diego, pp. 367-389.

Byers, JA (1995). Host-tree chemistry affecting colonization in bark beetles. In: Cardé, RT, Bell, WJ (eds), Chemical Ecology of Insects 2. Chapman and Hall, London; New York, pp. 154-213.

Cervone, F, Castoria, R, Leckie, F, De Lorenzo, G (1997). Perception of fungal elicitors and signal transduction. In: Aducci, P (ed), Signal Transduction in Plants. Birkhäuser, Basel, pp. 153-177.

Chappell, J (1995). Biochemistry and molecular biology of isoprenoid biosynthetic pathway in plants. Annu. Rev. Physiol. Plant Mol. Biol. 46:521-547.

Cheniclet, C (1987). Effects of wounding and fungus inoculation on terpene producing systems of maritime pine. J. Exp. Zool. 38:1557-1572.

Cipollini, D, Purrington, CB, Bergelson, J (2003). Costs of induced responses in plants. Basic Appl. Ecol. 4 :79-85.

Clancy, KM (2002). Mechanisms of resistance in trees to defoliators. In: Wagner, M, Clancy, K, Lieutier, F, Paine, T (eds), Mechanisms and Deployment of Resistance in Trees to Insects. Kluwer Academic Publishers, Dordrecht, pp. 79-103.

Codella, SG, Raffa, KF (2002). Desiccation of *Pinus* foliage induced by conifer sawfly oviposition: effect on egg viability. Ecol. Entomol. 27:618-621.

Codella, SG, Raffa, KF (1995a). Host plant influence on chemical defense in conifer sawflies (Hymenoptera: Diprionidae). Oecologia 104:1-11.

Codella, SG, Raffa, KF (1995b). Contributions of female oviposition pattern and larval behavior to group defense in conifer sawflies (Hymenoptera: Diprionidae). Oecologia 103:24-33.

Codella, SG, Raffa, KF (1993). Defense strategies of folivorous sawflies. In: Wagner, M, Raffa, K (eds), Sawfly Life History - Adaptations to Woody Plants. Academic Press, San Diego, pp. 261-294.

Coppel, HC, Benjamin, DM (1965). Bionomics of the nearctic pine-feeding diprionids. Annu. Rev. Entomol. 10:69-96.

Coppen, JJW, James, DJ, Robinson, JM, Subansenee, W (1998). Variability in xylem resin composition amongst natural populations of thai and filipino *Pinus merkusii* de Vriese. Flav. Frag. J. 13:33-39.

Croteau, R (1987). Biosynthesis and catabolism of monoterpenoids. Chem. Rev. 87:929-954.

de Groot, P, Turgeon, JJ (1998). Insect-pine interactions. In: Richardson, D (ed), Ecology and Biogeography of *Pinus*. Cambridge University Press, Cambridge, pp. 354-380.

Dicke, M (1999). Specificity of herbivore-induced plant defences. In: Chadwick DJ, Goode JA (eds), Insect-Plant Interactions and Induced Plant Defence. Wiley & Sons, Chichester, pp. 43-54.

Dicke, M, van Poecke, RMP (2002). Signalling in plant-insect interactions: signal transduction in direct and indirect plant defence. In: Scheel, D, Wasternack, C (eds), Plant Signal Transduction. Oxford University Press, Oxford, pp. 289-316.

Dicke, M, van Poecke, RMP, de Boer, JG (2003). Inducible indirect defence of plants: from mechanisms to ecological functions. Basic Appl. Ecol. 4:27-42.

Ebel, J, Mithöfer, A (1998). Early events in the elicitation of plant defence. Planta 206:335-348.

Eisner, T, Johnessee, JS, Carrel, J, Hendry, LB, Meinwald, J (1974). Defensive use by an insect of a plant resin. Science 184:996-999.

Erbilgin, N, Raffa, KF (2001a). Modulation of predator attraction to pheromones of two prey species by stereochemistry of plant volatiles. Oecologia 127:444-453.

Erbilgin, N, Raffa, KF (2001b). Kairomonal range of generalist predators in specialized habitats: responses to multiple phloeophagous species emitting pheromones vs. host odors. Entomol. Exp. Appl. 99:205-210.

Erbilgin, N, Raffa, KF (2000). Opposing effects of host monoterpenes on responses by two sympatric species of bark beetles to their aggregation pheromones. J. Chem. Ecol. 26:2527-2548.

Erbilgin, N, Powell, JS, Raffa, KF (2003). Effect of varying monoterpene concentration on the response of Ips pini (Coleoptera: Scolytidae) to its aggregation pheromone: implications for pest management and ecology of bark beetles. Agr. Forest Entomol. 5:269-274.

Fahn, A (1979). Secretory Structures in Plants. Academic Press, San Diego.

Fahn, A, Zamski, E (1970). The influence of pressure, wind, wounding and growth substances on the rate of resin duct formation in Pinus halepensis wood. Isr. J. Bot. 19:429-446.

Fäldt, J, Sjödin, K, Persson, M, Valterová, I, Borg-Karlson, A-K (2001). Correlation between selected monoterpene hydrocarbons in the xylem of six Pinus (Pinaceae) species. Chemoecology 11:97-106.

Feeny, P (1992). The evolution of chemical ecology: contributions from the study of insects. In: Rosenthal, G, Berenbaum, M (eds), Herbivores: Their Interactions with Secondary Plant Metabolites. VOl II. Ecological and Evolutionary Processes. Academic Press, San Diego, pp. 1-44.

Forrest, I, Burg, K, Klumpp, R (2000). Genetic markers: tools for identifying and characterising Scots pine populations. Invest. Agr. : Sist. Recur. For.: Fuera de Serie n. 1:67-88.

Franceschi, VR, Krokene, P, Krekling, T, Christiansen, E (2000). Phloem parenchyma cells are involved in local and distant defense responses to fungal inoculation or bark-beetle attack in Norway spruce (Pinaceae). Am. J. Bot. 87:314-326.

Gatehouse, JA (2002). Plant resistance towards insect herbivores: a dynamic interaction. New Phytol. 156:145-169.

Géri, C, Allais, J-P, Auger, M-A (1993). Effects of plant chemistry and phenology on sawfly behavior and development. In: Wagner, M, Raffa, KF (eds), Sawfly Life History Adaptations to Woody Plants. Academic Press, San Diego, pp. 173-210.

Gershenzon, J (1994). Metabolic costs of terpenoid accumulation in higher plants. J. Chem. Ecol. 20:1281-1328.

Gershenzon, J, Croteau, R (1991). Terpenoids. In: Rosenthal, G, Berenbaum, M (eds), Herbivores - Their Interactions with Secondary Metabolites Vol. 1 The Chemical Participants. Academic Press, San Diego, pp. 165-219.

Gijzen, M, Lewinsohn, E, Savage, TJ, Croteau, R (1993). Conifer monoterpenes. In: Teranishi, R, Buttery, R, Sugisawa, H (eds), Bioactive Volatile Compounds from Plants. ACS Symposium Series 525, Washington, DC., pp. 8-22.

Gleizes, M, Carde, J-P, Pauly, G, Bernard-Dagan, C (1980). In vivo formation of sesquiterpene hydrocarbons on the endoplasmatic reticulum of pine. Plant Sci. Lett. 20:79-90.

Gouinguené, S, Alborn, H, Turlings, TCJ (2003). Induction of volatile emissions in maize by different larval instars of Spodoptera littoralis. J. Chem. Ecol. 29:145-162.

Hanover, JW (1975). Physiology of tree resistance to insects. Annu. Rev. Entomol. 20:75-95.

Hare, JD (2002). Plant genetic variation in tritrophic interactions . In: Tscharntke, T, Hawkins, B (eds), Multitrophic Level Interactions. Cambridge University Press, Cambridge, pp. 8-43.

Haukioja, E, Honkanen, T (1997). Herbivore-induced responses in trees: internal vs. external explanations. In: Watt, A, Stork, N, Hunter, M (eds), Forests and Insects. Chapman & Hall, London, pp. 69-80.

Heidger, CM, Lieutier, F (2002). Possibilities to utilize tree resistance to insects on forest pest management in central and western Europe. In: Wagner, M, Clancy, K, Lieutier, F, Paine, T (eds), Mechanisms and Deployment of Resistance in Trees to Insects. Kluwer Academic Publishers, Dordrecht, pp. 239-263.

Hilker, M, Meiners, T (2002). Induction of plant responses towards oviposition and feeding of herbivorous arthropods: a comparison. Entomol. Exp. Appl. 104:181-192.

Hilker, M, Weitzel, C (1991). Oviposition deterrence by chemical signals of conspecific larvae in *Diprion pini* (Hymenoptera: Diprionidae) and *Phyllodecta vulgatissima* (Coleoptera: Chrysomelidae). Entomol. Gen. 15:293-301.

Hilker, M, Kobs, C, Varama, M, Schrank, K (2002). Insect egg deposition induces *Pinus* to attract egg parasitoids. J. Exp. Biol. 205:455-461.

Hilker, M, Bläske, V, Kobs, C, Dippel, C (2000). Kairomonal effects of sawfly sex pheromones on egg parasitoids. J. Chem. Ecol. 26:2591-2601.

Hiltunen, R, Tigerstedt, PMA, Juvonen, S, Pohjola, J (1975). Inheritance of 3-carene quantity in *Pinus sylvestris* L. Farm. Aikak. 84:69-72.

Karban, R, Baldwin, IT (1997). Induced responses to herbivory. Chicago University Press, Chicago.

Karban, R, Myers, JH (1989). Induced plant responses to herbivory. Annu. Rev. Ecol. Syst. 20:331-348.

Keely, JE, Zedler, PH (1998). Evolution of life histories in *Pinus*. In: Richardson, D (ed), Ecology and Biogeography in *Pinus*. Cambridge Academic Press, Cambridge, pp. 219-251.

Kleinhentz, M, Jactel, H, Menassieu, P (1999). Terpene attractant candidates of *Dioryctria sylvestrella* in maritime pine (*Pinus pinaster*) oleoresin, needles, liber, and headspace samples. J. Chem. Ecol. 25:2741-2756.

Knerer, G, Atwood, CE (1973). Diprionid sawflies: polymorphism and speciation. Science 179:1090-1099.

Komenda, M (2001). Investigations of the emissions of monoterpenes from Scots pine. PhD-Thesis University of Cologne.

Koricheva, J (2004). Meta-analysis of trade-offs among plant antiherbivore defenses: are plants jacks-of-all-trades, masters of all? Am. Nat. 163:E64-E75.

Krause, SC, Raffa, KF (1995). Defoliation intensity and larval age interact to affect sawfly performance on previously injured *Pinus resinosa*. Oecologia 102:24-30.

Kulman, HM (1971). Effects of insect defoliation on growth and mortality of trees. Annu. Rev. Entomol. 16:289-324.

Langenheim, JH (1994). Higher plant terpenoids: a phytocentric overview of their ecological roles. J. Chem. Ecol. 20:1223-1280.

Langenheim, JH (2003). Plant resins. Chemistry, Evolution, Ecology, and Ethnobotany. Timber Press, Portland.

Larsson, S (2002). Resistance in trees to insects - an overview of mechanisms and interactions. In: Wagner, M, Clancy, K, Lieutier, F, Paine, T (eds), Mechanisms and Deployment of Resistance in Trees to Insects. Kluwer Academic Publishers, Dordrecht, pp. 1-29.

Larsson, S, Ekbom, B, Björkman, C (2000). Influence of plant quality on pine sawfly population dynamics. Oikos 89:440-450.

Larsson, S, Björkman, C, Kidd, NAC (1993). Outbreaks in diprionid sawflies: why some species and not others. In: Wagner, M, Raffa, KF (eds), Sawfly Life History Adaptations to Woody Plants. Academic Press, San Diego, pp. 453-483.

Larsson, S, Björkman, C, Gref, R (1986). Responses of *Neodiprion sertifer* (Hym., Diprionidae) larvae to variation in needle resin acid concentration in Scots pine. Oecologia 70:77-84.

Latta, RG, Linhart, YA, Snyder, MA, Lundquist, L (2003). Patterns of variation and correlation in the monoterpene composition of xylem oleoresin within populations of ponderosa pine. Biochem. Syst. Ecol. 31:451-465.

Latta, RG, Linhart, YB, Lundquist, L, Snyder, MA (2000). Patterns of monoterpene variation within individual trees in ponderosa pine . J. Chem. Ecol. 26:1341-1357.

Lerdau, M, Gray, D (2003). Ecology and evolution of light-dependent and light-independent phytogenic volatile organic carbon. New Phytol. 157:199-211.

Lewinsohn, E, Gijzen, M, Croteau, R (1991a). Defense mechanisms of conifers- Differences in constitutive and wound-induced monoterpene biosynthesis among species. Plant Physiol. 96:44-49.

Lewinsohn, E, Gijzen, M, Savage, TJ, Croteau, R (1991b). Defensive mechanisms in conifers- Relationships of monoterpene cyclase activity to anatomical specialization and oleoresin monoterpene content. Plant Physiol. 96:38-43.

Lichtenthaler, HK (2000). Sterols and Isoprenoids. Biochem. Soc. Tans. 28:785-789.

Lichtenthaler, HK (1999). The -deoxy-D-xylulose-5-phosphate pathway of isoprenoid biosynthesis in plants. Annu. Rev. Physiol. Plant Mol. Biol. 50:47-65.

Lichtenthaler, HK, Schwender, J, Disch, A, Rohmer, M (1997). Biosynthesis of isoprenoids in higher plant chloroplasts proceeds via a mevalonate-independent pathway. FEBS Letters 400:271-274.

Lieutier, F (2002). Mechanisms of resistance in conifers and bark beetle attack strategies. In: Wagner, M, Clancy, K, Lieutier, F, Paine, T (eds), Mechanisms and Deployment of Resistance in Trees to Insects. Kluwer Academic Publishers, Dordrecht, pp. 31-77.

Lill, JT, Marquis, RJ, Ricklefs, RE (2002). Host plants influence parasitism of forest caterpillars. Nature 417:170-173.

Litvak, ME, Monson, RK (1998). Patterns of induced and constitutive monoterpene production in conifer needles in relation to insect herbivory. Oecologia 114:531-540.

Lombardero, MJ, Ayres, MP, Lorio, PL, Ruel, JJ (2000). Environmental effects on constitutive and inducible resin defences of *Pinus taeda*. Ecol. Letters 3:329-339.

Lyytikäinen, P (1994). The success of pine sawflies in relation to 3-carene and nutrient content of Scots pine foliage. For. Ecol. Man. 67:1-10.

Lyytikäinen, P (1992). Comparison of the effects of artificial and natural defoliation on the growth of diprionid sawflies on Scots pine foliage. J. Appl. Ent. 114:57-66.

Macchioni, F, Cioni, PL, Flamini, G, Maccioni, S, Ansaldi, M (2003). Chemical composition of essential oil from needles, branches and cones of *Pinus pinea*, *P. halepensis*, *P. pinaster*, and *P. nigra* from central Italy. Flavour Fragr. J. 18:139-143.

Manninen, A-M, Tarhanen, S, Vuorinen, M, Kainulainen, P (2002). Comparing the variation of needle and wood terpenoids in Scots pine provenances. J. Chem. Ecol. 28:211-228.

Martin, DM, Gershenzon, J, Bohlmann, J (2003). Induction of volatile terpene biosynthesis and diurnal emission by methyl jasmonate in foliage of Norway spruce. Plant Physiol. 132:1586-1599.

Mason, ME, Davis, JM (1997). Defense responses in slash pine: chitosan treatment alters the abundance of specific mRNAs. MPMI 10:135-137.

Mattiacci, L, Dicke, M, Posthumus, MA (1995). β–Glucosidase: An elicitor of herbivore-induced plant odor that attracts host-searching parasitic wasps. Proc. Natl. Acad. Sci. USA 92:2036-2040.

McCullough, DG, Wagner, MR (1993). Defusing host defenses: ovipositional adaptations of sawflies to plant resins. In: Wagner MR, Raffa KF (eds), Sawfly Life History Adaptations to Woody Plants. Academic Press, San Diego, pp. 157-172.

McKay, SAB, Hunter, W, Godard, K-A, Wang, SX, Martin, DM, Bohlmann, J, Plant, AL (2003). Insect attack and wounding induce traumatic resin duct development and gene expression of (-)-pinene synthase in Sitka spruce. Plant Physiol. 133:368-378.

Meiners, T, Hilker, M (2000). Induction of plant synomones by oviposition of a phytophagous insect. J. Chem. Ecol. 26:221-232.

Miller, DR, Borden, JH (1990). β–Phellandrene: kairomone for pine engraver *Ips pini* (Say) (Coleoptera: Scolytidae). J. Chem. Ecol. 16:2519-2531.

Miller, DR, Borden, JH (2000). Dose-dependent and species specific responses of the pine bark beetles (Coleoptera: Scolytidae) to monoterpenes in association with pheromones. Can. Entomol. 132:183-195.

Miller, RH, Berryman, AA, Ryan, CA (1986). Biotic elicitors of defense reactions in lodgepole pine. Phytochemistry 25:611-612.

Mirov, NT (1967). The Genus *Pinus*. Ronald Press Company, New York.

Mumm, R, Schrank, K, Wegener, R, Schulz, S, Hilker, M (2003). Chemical analysis of volatiles emitted by *Pinus sylvestris* after induction by insect oviposition. J. Chem. Ecol. 29:1235-1252.

Nagy, NE, Franceschi, VR, Solheim, H, Krekling, T, Christiansen, E (2000). Wound-induced traumatic resin duct development in stems of Norway spruce (Pinaceae): Anatomy and cytochemical traits. Am. J. Bot. 87:302-313.

Nerg, A-M, Heijari, J, Noldt, U, Viitanen, H, Vuorinen, M, Kainulainen, P, Holopainen, JK (2004). Significance of wood terpenoids on the resistance of Scots pine provenances against the old house borer, *Hylotrupes bajulus*, and brown-rot fungus, *Coniophora puteana*. J. Chem. Ecol. 30:125-141.

Nerg, A, Kainulainen, P, Vuorinen, M, Hanso, M, Holopainen, JK, Kurkela, T (1994). Seasonal and geographical variation of terpenes, resin acids and total phenolics in nursery grown seedlings of Scots pine (*Pinus sylvestris* L.). New Phytol. 128:703-713.

Niemelä, P, Tuomi, J, Lojander, T (1991). Defoliation of the Scots pine and performance of diprionid sawflies. J. Anim. Ecol. 60:683-692.

Niemelä, P, Tuomi, J, Mannila, R, Ojala, P (1984). The effect of previous damage on the quality of Scots pine foliage as food for diprionid sawflies. Z. ang. Ent. 98:33-43.

Nordlander, G (1990). Limonene inhibits attraction to α-pinene in the pine weevils *Hylobius abietis* and *H. pinastri*. J. Chem. Ecol. 16:1307-1320.

Nykänen, H, Koricheva, J (2004). Damage induced changes in woody plants and their effects on insect herbivore performance: a meta-analysis. Oikos 104:247-268.

Olofsson, E (1989). Oviposition behaviour and host selection on *Neodiprion sertifer* (Geoffr.) (Hym., Diprionidae). J. Appl. Entomol. 107:357-364.

Paine, TD, Raffa, KF, Harrington, TC (1997). Interactions among scolytid bark beetles, their associated fungi, and live host conifers. Annu. Rev. Entomol. 42:179-206.

Park, KC, Zhu, J, Harris, J, Ochieng, SA, Baker, TC (2001). Electroantennogram responses of parasitic wasp, *Microplitis croceipes*, to host-related volatile and anthropogenic compounds. Physiol. Entomol. 26:69-77.

Pasquier-Barré, F, Palasse, C, Goussard, F, Auger-Rozenberg, M-A, Géri, C (2001). Relationship of Scots pine clone characteristics and water stress to hatching and larval performance of the sawfly *Diprion pini* (Hymenoptera: Diprionidae). Environ. Entomol. 30:1-6.

Pasquier-Barré, F, Geri, C, Goussard, F, Auger-Rozenberg, M-A, Grenier, S (2000). Oviposition preference and larval survival of *Diprion pini* on Scots clones in relation to foliage characteristics. Agr. For. Entomol. 2:185-192.

Petrakis, PV, Tsitsimpikou, C, Tzakou, O, Couladis, M, Vagias, C, Roussis, V (2001). Needle volatiles from five *Pinus* species growing in Greece. Flavour Frag. J. 16:249-252.

Phillips, MA, Croteau, R (1999). Resin-based defenses in conifers. Trends Plant Sci. 4:184-190.

Phillips, MA, Wildung, MR, Wialliams, DC, Hyatt, DC, Croteau, R (2003). cDNA isolation, functional expression, and characterization of (+)-α-pinene synthase from loblolly pine (*Pinus taeda*). Stereocontrol on pine biosynthesis. Arch. Biochem. Biophys. 411:267-276.

Phillips, MA, Savage, TJ, Croteau, R (1999). Monoterpene synthases of loblolly pine (*Pinus taeda*) produce pinene isomeres and enantiomeres. Arch. Biochem. Biophys. 372:197-204.

Popp, MP, Johnson, JD, Lesney, MS (1995). Characterization of the induced response of slash pine to inoculation with bark beetle vectored fungi. Tree Physiol. 15:619-623.

Raffa, KF (2001). Mixed messages across multiple trophic levels: the ecology of bark beetle chemical communication systems. Chemoecology 11:49-65.

Raffa, KF (1991). Induced defensive reactions in conifer-bark beetle systems. In: Tallamy, D, Raupp, M (eds), Phytochemical Induction by Herbivores. John Wiley & Sons, New York, pp. 245-276.

Raffa, KF, Dahlsten, DL (1995). Differential responses among natural enemies and prey to bark beetle pheromones. Oecologia 102:17-23.

Raffa, KF, Smalley, EB (1995). Interaction of pre-attack and induced monoterpene concentrations in host conifer defense against bark beetle-fungal complexes. Oecologia 102:285-295.

Raffa, KF, Berryman, AA (1987). Interacting selective pressures in conifer-bark beetle systems: A basis for reciprocal adaptations. Am. Nat. 129:234-262.

Raffa, KF, Krause, SC, Reich, PB (1998). Long-term effects of defoliation on red pine suitability to insects feeding on diverse plant tissues. Ecology 79:2352-2364.

Rafii, ZA, Dodd, RS, Zavarin, E (1996). Genetic diversity in foliar terpenoids among natural populations of European Black pine. Biochem. Syst. Ecol. 24:325-339.

Richardson, DM, Rundel, PW (1998). Ecology and biogeography of *Pinus*: an introduction. In: Richardson, D (ed), Ecology and Biogeography of *Pinus*. Cambridge University Press, Cambridge, pp. 3-46.

Roberts, DR (1970). Within-tree variation of monoterpene hydrocarbon composition of slash pine oleoresin. Phytochemistry 9:809-815.

Roussis, V, Petrakis, PV, Ortiz, A, Mazomenos, BE (1995). Volatile constituents of needles of five *Pinus* species grown in Greece. Phytochemistry 39:357-361.

Röse, U, Tumlinson, JH (2004). Volatiles released from cotton plants in response to *Helicoverpa zea* feeding damage on cotton flower buds. Planta 218:824-832.

Sadof, CS, Grant, GG (1997). Monoterpene composition of *Pinus sylvestris* varieties and susceptible to *Dioryctria zimmermani*. J. Chem. Ecol. 23:1917-1927.

Saikkonen, K, Neuvonen, S, Kainulainen, P (1995). Oviposition and larval performance of European pine sawfly in relation to irrigation, simulated acid rain and resin acid concentration in Scots pine. Oikos 74:273-282.

Salin, F, Pauly, G, Charon, J, Gleizes, M (1995). Purification and characterization of trans-β-farnesene synthase from maritime pine (*Pinus pinaster* Ait.) needles. J. Plant Physiol. 146:203-209.

Sallas, L, Kainulainen, P, Utriainen, J, Holopainen, T, Holopainen, JK (2001). The influence of elevated $O_3$ and $CO_2$ concentrations on secondary metabolites of Scots pine (*Pinus sylvestris* L.) seedlings. Global Change Biol. 7:303-311.

Savage, TJ, Ichii, H, Hume, SD, Little, DB, Croteau, R (1995). Monoterpene synthases from gymnosperms and angiosperms: stereospecificity and activation by cysteinyl- and arginyl modifying reagents. Arch. Biochem. Biophys. 320:257-265.

Savage, TJ, Hatch, MW, Croteau, R (1994). Monoterpene synthases of *Pinus contorta* and related conifers. J. Biol. Chem. 269:4012-4020.

Schwerdtfeger, F (1981). Waldkrankheiten. Parey, Hamburg.

Shao, M, Czapiewski, KV, Heiden, AC, Kobel, K, Komenda, M, Koppmann, R, Wildt, J (2001). Volatile organic compound emission from Scots pine: mechanisms and description by algorithms. J. Geophys. Res. 106:20483-20491.

Shu, S, Grant, GG, Langevin, D, Lombardo, DA, MacDonald, L (1997). Oviposition and electroantennogram responses of *Dioryctria abietivorella* (Lepidoptera: Pyralidae) elicited by monoterpenes and enantiomeres from eastern white pine. J. Chem. Ecol. 23:35-50.

Sitte, P, Ziegler, H, Ehrendorfer, F, Bresinsky, A (1991). Strasburger - Lehrbuch der Botanik. Gustav Fischer, Stuttgart.

Sjödin, K, Persson, M, Fäldt, J, Ekberg, I, Borg-Karlson, A-K (2000). Occurrence and correlations of monoterpene hydrocarbon enantiomers in *Pinus sylvestris* and *Picea abies*. J. Chem. Ecol. 26:1701-1720.

Sjödin, K, Persson, M, Borg-Karlson, A-K, Norin, T (1996). Enantiomeric compositions of monoterpene hydrocarbons in different tissues of four individuals of *Pinus sylvestris*. Phytochemistry 41:439-445.

Sjödin, K, Schroeder, LM, Eidmann, HH, Norin, T, Wold, S (1989). Attack rates of scolytids and composition of volatile wood constituents in healthy and mechanically weakened pine trees. Scand. J. For. Res. 4:379-391.

Smits, A, Larsson, S (1999). Effects of previous defoliation on pine looper larval performance. Agr. Forest Entomol. 1:19-26.

Stiling, P, Simberloff, D (1989). Leaf abscission: induced defense against pests or response to damage? Oikos 55:43-49.

Sullivan, BT, Berisford, CW (2004). Semiochemicals from fungal associates of bark beetles may mediate host location behavior of parasitoids. J. Chem. Ecol. 30:703-717.

Sullivan, BT, Pettersson, EM, Seltmann, KC, Berisford, CW (2000). Attraction of the bark beetle parasitoid *Roptrocerus xylophagorum* (Hymenoptera: Pteromalidae) to host associated olfactory cues. Environ. Entomol. 29:1138-1151.

Tiberi, R, Niccoli, A, Curini, M, Epifano, F, Marcotullio, MC, Rosati, O (1999). The role of the monoterpene composition in *Pinus* spp. needles, in host selection by the pine processionary caterpillar, *Thaumetopoea pityocampa*. Phytoparasitica 27:263-272.

Tingey, DT, Turner, DP, Weber, JA (1991). Factors controlling the emissions of monoterpenes and other volatile organic compounds. In: Sharkey TD, Holland EA, Mooney HA (eds), Trace Gas Emissions by Plants. Academic Press, San Diego, pp. 93-119.

Tomlin, ES, Alfaro, RI, Borden, JH, He, F (1998). Histological response of resistant and susceptible white spruce to simulated white pine weevil damage. Tree Physiol. 18:21-28.

Trapp, S, Croteau, R (2001). Defensive resin biosynthesis in conifers. Annu. Rev. Plant Physiol. Plant Mol. Biol. 52:589-724.

Trewhalla, KE, Leather, SR, Day, KR (2000). Variation in the suitability of *Pinus contorta* (lodgepole pine) to feeding by three pine defoliators, *Panolis flammea*, *Neodiprion sertifer*, and *Zeiraphea diniana*. J. Appl. Ent. 124:11-17.

Turlings, TCJ, Gouinguené, S, Degen, T, Fritzsche-Hoballah, ME (2002). The chemical ecology of plant-caterpillar-parasitoid interactions. In: Tscharntke, T, Hawkins, B (eds), Multitrophic Level Interactions. Cambridge University Press, Cambridge, pp. 148-173.

Turlings, TCJ, Bernasconi, M, Bertossa, R, Bigler, F, Caloz, G, Dorn, S (1998). The induction of volatile emissions in maize by three herbivore species with different feeding habitats: possible consequences for their natural enemies. Biol. Control 11:122-129.

Turlings, TCJ, Wäckers, FL, Vet, LEM, Lewis, WJ, Tumlinson, JH (1993). Learning of host-finding cues by hymenopterous parasitoids. In: Papaj, DR, Lewis, AC (eds), Insect Learning. Ecological and Evolutionary Perspectives. Chapman & Hall, New York, pp. 51-78.

Turlings, TCJ, Tumlinson, JH, Heath, RR, Proveaus, AT, Doolittle, RE (1991). Isolation and identification of allelochemicals that attract the larval parasitoid, *Cotesia marginiventris* (Cresson), to the microhabitat of one of its hosts. J. Chem. Ecol. 17:2235-2251.

Turlings, TCJ, Tumlinson, JH, Lewis, WJ (1990). Exploitation of herbivore-induced plant odors by host-seeking parasitic wasps. Science 250:1251-1253.

Turtola, S, Manninen, A-M, Rikala, R, Kainulainen, P (2003). Drought stress alters the concentration of wood terpenoids in Scots pine and Norway spruce seedlings. J. Chem. Ecol. 29:1981-1995.

Valterová, I, Sjödin, K, Norin, T, Norin, T (1995). Contents and enantiomeric compositions of monoterpene hydrocarbons in xylem oleoresins from four *Pinus* species growing in Cuba. Comparison of trees unattacked and attacked by *Dioryctria horneana*. Biochem. Syst. Ecol. 23:1-15.

van den Boom, CEM, van Beek, TA, Posthumus, MA, de Groot, A, Dicke, M (2004). Qualitative and quantitative variation among volatile profiles induced by *Tetranychus urticae* feeding on plants from various families. J. Chem. Ecol. 30:69-89.

Vet, LEM, Dicke, M (1992). Ecology of infochemical use by natural enemies in a tritrophic context. Annu. Rev. Entomol. 37:141-172.

Vet, LEM, Lewis, WJ, Cardé, RT (1995). Parasitoid foraging and learning. In: Cardé, RT, Bell, WJ (eds), Chemical Ecology of Insects 2. Chapman and Hall, London; New York, pp. 65-101.

Völkl, W (2000). Foraging behaviour and sequential multisensory orientation in the aphid parasitoid, *Pauesia picta* (Hym., Aphidiidae) at different spatial scales. J. Appl. Entomol. 124:307-314.

Wagner, MR, Zhang, Z (1993). Host plant traits associated with resistance of ponderosa pine to the sawfly, *Neodiprion fulviceps*. Can. J. For. Res. 23:839-845.

Wagner, MR, Raffa, KF (1993). Sawfly Life History Adaptations to Woody Plants. Academic Press, San Diego.

Wainhouse, D, Cross, DJ, Howell, RS (1990). The role of lignin as a defence against the spruce bark beetle *Dendroctonus micans*: effect on larvae and adults. Oecologia 85:257-265.

Wallin, KF, Raffa, KF (1999). Altered constitutive and inducible phloem monoterpenes following natural defoliation of Jack pine: implications to host mediated interguild interactions and plant defense theories. J. Chem. Ecol. 25:861-880.

Weitzel C. (1991). Eiablageregulierende Ökomone bei ausgewählten phytophagen Insekten: *Phyllodecta vulgatissima* L. (Coleoptera, Chrysomelidae), *Diprion pini* (L.) und *Gilpinia hercyniae* Htg. (Hymenoptera, Diprionidae). PhD Thesis University of Bayreuth.

Werker, E, Fahn, A (1969). Resin ducts of *Pinus halepensis* Mill.-Their structure, development and pattern of arrangement. Bot. J. Lin. Soc. 62:379-411.

Wibe, A, Borg-Karlson, A-K, Persson, M, Norin, T, Mustaparta, H (1998). Enantiomeric composition of monoterpene hydrocarbons in some conifers and receptor neuron discrimination of α-pinene and limonene enantiomers in the pine weevil, *Hylobius abietis*. J. Chem. Ecol. 24:273-287.

Williams, AG, Whitham, TG (1986). Premature leaf abscission: an induced plant defense against gall aphids. Ecology 67:1619-1627.

Wu, H, Hu, Z-H (1997). Comparative anatomy of resin ducts of the Pinaceae. Trees 11:135-143.

Zhang, Q-H, Schlyter, F, Battisti, A, Birgersson, G, Anderson, P. (2003). Electrophysiological responses of *Thaumetopoea pityocampa* females to host volatiles: implications for host selection of active and inactive terpenes. J. Pest Science 76:103-107.

# Chapter 8

# Summary

**Bold: most important results and conclusions**

<u>Underlined:</u> methods used

Plants may defend themselves when attacked by herbivores. They are not just able to mobilize their defence when attacked by feeding herbivores, but may even "notice" egg deposition by insects. Plants have evolved several defensive responses to insect egg deposition (Hilker et al. 2002b). One of these responses is to produce volatiles which attract egg parasitoids killing the eggs. With this preventive defensive strategy a plant is able to escape from herbivory prior to hatching of hungry larvae. The PhD study presented here investigated the mechanisms and functions of the response of Scots pine (*Pinus sylvestris* L.) to egg deposition by the pine sawfly *Diprion pini* L. (Hymenoptera, Diprionidae).

In the beginning of this PhD study it was known that egg deposition by this sawfly induces pine to emit volatiles attracting the egg parasitoid *Chrysonotomyia ruforum* Krausse (Hymenoptera, Eulophidae). The pine´s response to *D. pini* egg deposition was known to act both on a local scale (i.e. at the pine twig carrying sawfly eggs) and a systemic scale (i.e. an egg-free part of the pine twig adjacent to an egg-carrying one). Treatment of pine twigs with jasmonic acid resulted into the emission of volatiles which also attracted the egg parasitoids. Furthermore, the oviduct secretion coating the eggs was found to contain an elicitor inducing the pine's response to egg deposition. Application of the oviduct secretion into artificially wounded pine needles also resulted in the induction of volatiles attractive to the egg parasitoid. Artificially wounding mimicked the mechanical damage inflicted by the ovipositor of the female sawfly during egg deposition. Volatiles released by artificially wounded pines were shown to be not attractive to the egg parasitoids (Hilker et al. 2002a) (compare *Chapter 1*).

Within this PhD study, further experiments were conducted to investigate the chemical characteristics and stability of the elicitor. **Wounding of pine needles was essential for the induction process since the elicitor was only active when transferred into damaged needle tissue.** The activity of the elicitor was not stable and got lost, when the secretion was stored or frozen in Aqua dest. However, the activity of the secretion was kept at room temperature and even at –80 °C, when oviduct secretion was diluted in Ringer solution. The activity of the elicitor in the oviduct secretion vanished after treatment of the oviduct secretion with proteinase K, which destroyed all proteins, as was shown by <u>SDS-PAGE</u>. **These results indicate that the elicitor in the oviduct secretion is a peptide or protein, or a component bound to these** (*Chapter 2*).

In order to elucidate the pine volatiles induced by egg deposition, the headspace of systemically oviposition-induced twigs of *P. sylvestris* was compared with volatiles from artificially wounded (control) twigs by <u>gas chromatography – mass spectrometry (GC-MS)</u> analyses. In addition, the parasitoid-attracting volatile blend emitted by JA-treated pine twigs was compared with odour of untreated controls. Pine twigs emitted the same mono- and sesquiterpenoids regardless of their treatment. **Thus, no qualitative change in the headspace of pine twigs was detected after oviposition or JA-treatment. A quantitative comparison revealed that only the sesquiterpene (*E*)-β-farnesene was emitted in significant higher amounts by oviposition- and JA-induced twigs compared to the respective controls** (*Chapter 3*).

Subsequent experiments examined whether (*E*)-β-farnesene is utilized by *C. ruforum* as chemical cue to locate pine infested with eggs of *D. pini*. The behavioural response of *C. ruforum* to different concentrations of (*E*)-β-farnesene was tested by using <u>olfactometer bioassays</u>. The egg parasitoid did not respond to this sesquiterpene at neither concentration tested. However, egg parasitoids responded significantly to (*E*)-β-farnesene when this compound was offered in combination with a background odour consisting of the natural volatile blend emitted from a pine twig without eggs. This response was dependent on the applied concentration of $(E)$-β-farnesene. Only the concentration of 100ng/μl was able to attract *C. ruforum* females, whereas the concentration of 10ng/μl $(E)$-β-farnesene was not. Other sesquiterpenes $((E)$-β-caryophyllene, δ-cadinene) of the odour blend of pine that were offered in combination with volatiles of egg-free pine as background odour were not attractive to the egg parasitoids. **Thus, background odour was essential for the egg parasitoid to respond behaviourally to $(E)$-β-farnesene. These experiments indicate that *C. ruforum* compares the ratio of specifically $(E)$-β-farnesene to a pine volatile background which might "tell" the egg parasitoid where to find host eggs** (*Chapter 4*).

Further, the specificity of chemical plant cues used by the egg parasitoid *C. ruforum* for locating host eggs was examined. In <u>olfactometer bioassays</u> it was also tested whether female egg parasitoids respond innately to oviposition-induced pine volatiles. Naïve *C. ruforum* females were not attracted to *P. sylvestris* volatiles induced by oviposition of *D. pini*, but they were able to learn to respond to oviposition-induced pine volatiles when previously having experienced a pine twig with host eggs. Plant specificity of this tritrophic interaction was investigated by testing the response of experienced *C. ruforum* females to volatiles of Austrian black pine (*Pinus nigra* Arnold var. *nigra*) carrying eggs of *D. pini*. Herbivore specificity was studied by testing the response of experienced *C. ruforum* to volatiles from *P. sylvestris* twigs on which eggs of *Gilpinia pallida* Klug or *Neodiprion sertifer* Geoffroy, two suitable host species of *C. ruforum*, were deposited. In addition, it was studied whether larval feeding of *D. pini* also induces volatiles in *P. sylvestris* that attract female *C. ruforum*.

Neither volatiles from *P. nigra* with eggs of *D. pini* nor volatiles from *P. sylvestris* carrying eggs of *G. pallida* were attractive to female *C. ruforum*, even though the parasitoids had experienced the same plant-host complex prior to the bioassay. In contrast, parasitoids were significantly attracted to volatiles from *P. sylvestris* induced by egg deposition of *N. sertifer*. Feeding by sawfly larvae did not induce volatiles attractive to the egg parasitoids.

**In conclusion, the egg parasitoid *C. ruforum* specialized on diprionid hosts was shown to be able to learn only those cues specific for the plant species that is most beneficial for herbivore performance, for the herbivore species most abundant, and for the developmental stage (i.e. the egg stage) suitable for parasitism. These results further showed that a specialist egg parasitoid like *C. ruforum* does not respond *innately* to cues specific for its host, but instead needs to learn them, thus showing phenotypic plasticity of a narrow frame. These results indicate the necessity to open the paradigm that a specialist parasitoid responds innately to plant-host cues, whereas a generalist needs to learn** (Steidle and van Loon 2003) (*Chapter 5*).

Since egg deposition of *D. pini* did not induce *Pinus nigra* to release volatiles attractive for *C. ruforum*, the volatiles of *P. nigra* twigs that were either untreated, artificially wounded, or laden with eggs of *D. pini* were analysed by <u>GC-MS</u>. Artificially wounding or egg deposition significantly affected the quantitative composition of the headspace of *P. nigra*. A comparison of the volatile compositions of artificially wounded and egg-carrying twigs of *P. nigra* on one hand and oviposition-induced and artificially wounded twigs of *P. sylvestris* on the other hand by <u>multivariate data analyses</u> revealed significant qualitative and quantitative differences. **The "non-response" of *C. ruforum* to volatiles of egg-carrying *P. nigra* twigs might be due several factors: (1) the wrong quantitative volatile composition, (2) the lack of attractive key components in the headspace of egg-carrying *P. nigra* or conversely, (3) mixture suppression by additionally emitted compounds which mask the attractiveness of key compounds in the volatile mixture** (*Chapter 6*).

Finally, the oviposition-induced indirect defence mechanisms of *P. sylvestris* are embedded in an overview of general chemical defence mechanisms of pines (*Chapter 7*). Pines produce copious amounts of viscous oleoresin constitutively, which in response to wounding is translocated and accumulated to the wounding site. In addition, the biosynthesis of resin is induced. Compared to other conifer genera, pines generally show weaker induced responses. The pine's response induced by egg deposition is not strong since significant changes in the volatile pattern were only detected for $(E)$-β-farnesene.

However, in this tritrophic system very fine-tuned and specific chemical interactions seem to have evolved. Egg parasitoids learn to respond specifically to $(E)$-β-farnesene, the most significantly oviposition-induced terpenoid in *P. sylvestris*, when this compound was experienced in the "right" context, i.e. with a background odour of non-oviposition-induced pine volatiles. On the other hand, pines respond specifically upon the egg deposition, likely mediated by the involvement of specific elicitors in the oviduct secretions of female sawflies.

**References**

Hilker, M, Meiners, T (2002). Induction of plant responses towards oviposition and feeding of herbivorous arthropods: a comparison. Entomol. Exp. Appl. 104:181-192.

Hilker, M, Kobs, C, Varama, M, Schrank, K (2002a). Insect egg deposition induces *Pinus* to attract egg parasitoids. J. Exp. Biol. 205:455-461.

Hilker, M, Rohfritsch, O, Meiners, T (2002b). The plant's response towards insect egg deposition. In: Hilker, M, Meiners, T (eds), Chemoecology of Insect Eggs and Egg Deposition. Blackwell Publishing, Berlin, Oxford, pp. 205-233.

Steidle, JLM, van Loon, JJA (2003). Dietary specialization and infochemical use in carnivorous arthropods: testing a concept. Entomol. Exp. Appl. 108:133-148.

# Chapter 9

# Zusammenfassung

**Fett:** die wichtigsten Ergebnisse und Schlussfolgerungen

Unterstrichen: verwendete Methoden

Pflanzen können sich gegen die Angriffe von Herbivoren verteidigen. Sie sind nicht nur in der Lage, ihre Verteidigung zu mobilisieren, wenn sie durch fressende Herbivore angegriffen werden, sondern „bemerken" bereits, wenn Eier auf ihnen abgelegt werden. Pflanzen haben verschiedene Verteidigungsreaktionen auf Eiablage herbivorer Insekten entwickelt (Hilker et al. 2002b). Eine Möglichkeit ist, flüchtige Duftstoffe zu produzieren, die Eiparasitoiden anlocken, welche die Eier töten. Mit dieser präventiven Verteidigungsstrategie kann eine Pflanze dem Fraß vor dem Schlupf hungriger Larven zu entkommen. In der vorliegenden Dissertation wurden die Mechanismen und Funktionen der Reaktion der Waldkiefer (*Pinus sylvestris* L.) auf die Eiablage der Kiefernbuschhorn-Blattwespe *Diprion pini* L. (Hymenoptera, Diprionidae) untersucht.

Zu Beginn dieser Arbeit war bekannt, dass die Eiablage dieser Blattwespe in der Kiefer die Freisetzung von flüchtigen Duftstoffen induziert, die den Eiparasitoiden *Chrysonotomyia ruforum* Krausse (Hymenoptera, Eulophidae) anlocken. Die Reaktion der Kiefer auf die Eiablage von *D. pini* ist sowohl lokal (d.h. an dem Kiefernzweig, an dem Eier abgelegt wurden) als auch systemisch (d.h. Teile eines Zweiges an dem Eier abgelegt wurden, die jedoch keine Eier enthielten). Eine Behandlung der Kiefernzweigen mit Jasmonsäure (JA) induzierte die Freisetzung von flüchtigen Duftstoffen, die auf die Eiparasitoiden attraktiv wirkten. Weiterhin wurde gezeigt, dass in dem Oviduktsekret, welches die Eier umhüllt, ein Elicitor enthalten ist, der die Reaktion der Kiefer auf die Eiablage induziert. Die Applikation des Oviduktsekrets in artifiziell verletzte Kiefernnadeln induzierte ebenfalls die Abgabe flüchtiger Düfte, die Eiparasitoiden anlockten. Die artifizielle Verwundung simulierte die mechanische Beschädigung, die durch den Legebohrer des

Blattwespenweibchen während der Eiablage entsteht. Düfte von artifiziell verletzten Kiefern waren nicht attraktiv für die Eiparasitoiden (Hilker et al. 2002a) (vergleiche *Kapitel 1*).

In dieser Arbeit wurden Experimente durchgeführt, um die chemischen Merkmale und die chemische Stabilität des Elicitors weiter zu untersuchen. **Für den Induktionsprozess ist die Verletzung der Kiefernnadeln notwendig, da der Elicitor nur dann aktiv war, wenn er in beschädigtes Nadelgewebe eingebracht wurde.** Die Aktivität des Elicitor war nicht stabil und ging verloren, wenn das Oviduktsekret in Aqua dest. aufbewahrt oder eingefroren wurde. Wurde das Oviduktsekret jedoch in Ringerlösung gelöst, so blieb die Aktivität des Elicitors auch nach längerem Aufbewahren bei Raumtemperatur oder bei −80 °C erhalten. Eine Behandlung des Oviduktsekrets mit Proteinase K, die alle Proteine verdaut, führte zum Aktivitätsverlust des Elicitors, was durch eine SDS-PAGE bestätigt wurde. **Diese Befunde deuten darauf hin, dass es sich bei dem Elicitor im Oviduktsekret um ein Peptid oder Protein handelt oder um eine Komponente, die an ein Protein gebunden ist** (*Kapitel 2*).

Um die flüchtigen Kieferndüfte, die durch die Eiablage induziert werden zu analysieren, wurden mittels gekoppelter Gaschromatographie-Massenspektrometrie (GC-MS) die in den sog. „headspace" abgegebenen flüchtigen Verbindungen von eiablage-induzierten *P. sylvestris* Zweigen mit denen von artifiziell verletzten Kiefernzweigen (Kontrolle) verglichen. Weiterhin wurden die, für Eiparasitoiden attraktive, Düfte von JA-induzierten Kiefernzweigen mit den Düften unbehandelter Kontrollen verglichen. Unabhängig von der vorangegangenen Behandlung gaben Kiefernzweige die gleichen Mono- und Sesquiterpenoide ab. **Es wurden daher keine qualitative Veränderungen im Duftmuster nach Eiablage oder JA-Behandlung gefunden. Ein quantitativer Vergleich ergab, dass eiablage- und JA-induzierte Zweige im Vergleich zu den jeweiligen Kontrollen lediglich das Sesquiterpen (*E*)-β-Farnesen in signifikant erhöhten Mengen abgegeben** (*Kapitel 3*).

In nachfolgenden Experimenten wurde untersucht, ob (E)-β-Farnesen von C. ruforum als chemisches Signal zum Auffinden von mit Eiern von D. pini befallenen Zweigen nutzt. In Biotests wurde die Verhaltensantwort von C. ruforum auf verschiedene Konzentrationen von (E)-β-Farnesen im Olfaktometer getestet. Die Eiparasitoiden zeigten auf keine der getesteten Konzentrationen dieses Sesquiterpenes eine signifikante Reaktion. Wenn den Eiparasitoiden jedoch (E)-β-Farnesen gemeinsam mit Düften eines unbefallenen Kiefernzweiges als Hintergrundduft angeboten wurde, dann zeigten sie eine signifikante Reaktion, die abhängig von der angebotenen Konzentration vom (E)-β-Farnesen war. Während die Konzentration von 100ng/µl attraktiv war, zeigte die Konzentration von 10ng/µl (E)-β-Farnesen keine attraktive Wirkung. Weitere Sesquiterpene ((E)-β-Caryophyllen, δ-Cadinen) aus dem Duftbouquet von Kiefern, die gemeinsam mit Düften unbefallener Kiefernzweige als Hintergrund angeboten wurde, waren für die Eiparasitoiden nicht attraktiv. **Somit ist ein Hintergrundduft essentiell, damit der Eiparasitoid eine Verhaltensreaktion auf (E)-β-Farnesen zeigt. Diese Experimente deuten darauf hin, dass C. ruforum das Verhältnis von speziell (E)-β-Farnesen zu einem Hintergrund bestehend aus Kieferndüften vergleicht und dies einen Hinweis darauf wo Wirtseier zu finden sind** (Kapitel 4).

Desweiteren wurde die Spezifität der Pflanzendüfte, die von C. ruforum zum Auffinden von Wirteiern genutzt werden, untersucht. Mit Biotests im Olfaktometer wurde außerdem getestet, ob weibliche Eiparasitoiden auch angeborenermaßen auf eiablage-induzierte Kieferndüfte reagieren. Naive C. ruforum wurden durch Düfte von P. sylvestris, die durch Eiablagen von D. pini induziert wurden, nicht angelockt. Die Reaktion auf eiablage-induzierte Kieferndüfte konnte von den Parasitoiden erlernt werden, wenn sie zuvor auf einem Kiefernzweig mit Wirtseiern Erfahrung sammeln konnten. Die Pflanzenspezifität bezüglich dieser tritrophischen Interaktionen wurde untersucht indem die Reaktion von erfahrenen C. ruforum Weibchen auf Düfte von Schwarzkiefern (Pinus nigra Arnold var. nigra) getestet wurde auf die D. pini zuvor Eier abgelegt hatte. Für die Herbivorenspezifität wurde die Reaktion von erfahrenen C. ruforum Weibchen auf Düfte von P. sylvestris getestet, auf die zuvor Eier von Gilpinia pallida Klug beziehungsweise Neodiprion *sertifer* Geoffroy, zwei Wirtsarten von *C. ruforum*, abgelegt wurden. Außerdem wurde geprüft, ob auch Larvalfraß von *D. pini* Düfte in *P. sylvestris* induzieren kann, die weibliche Eiparasitoiden anlocken.

Weder Düfte von *P. nigra* mit Eiern von *D. pini* noch Düfte von *P. sylvestris* mit Eiern von *G. pallida* waren attraktiv für weibliche *C. ruforum*, obwohl die Parasitoiden vor dem Biotests auf dem gleichen Pflanze-Wirts-Komplex Erfahrung gesammelt hatten. Im Gegensatz dazu wurden Parasitoiden signifikant von Düften von *P. sylvestris*, die durch Eiablagen von *N. sertifer* induziert wurden, angelockt. Fraß von Blattwespenlarven induzierte keine für Eiparasitoiden attraktiven Düfte.

**Zusammenfassend zeigen die Ergebnisse dieser Experimente, dass der auf Eier von Vertretern der Diprionidae spezialisierte Eiparasitoid *C. ruforum* nur solche Düfte lernt, die spezifisch sind, für die Pflanzenart, welche am geeignetsten für seinen Wirt ist, für die am häufigsten auftretenden Wirtsarten und für das Entwicklungsstadium (d.h. das Ei), dass für eine Parasitierung geeignet ist. Außerdem zeigen die Ergebnisse, dass ein spezialisierter Eiparasitoid wie *C. ruforum* keine angeborene Reaktion auf spezifische Wirtsdüfte zeigt, sondern diese erst erlernen muss und somit eine phänotypische Plastizität in einem engen Rahmen besitzt. Dies lässt die Notwendigkeit erkennen, das Paradigma, dass spezialisierte Parasitoiden angeborene Reaktionen auf Pflanze-Wirts-Signale zeigen, während Generalisten dies erst erlernen müssen, zu öffnen** (Steidle and van Loon 2003) (*Kapitel 5*).

Da *P. nigra* nicht durch die Eiablage von *D. pini* induziert wird Düfte abzugeben, die *C. ruforum* anlocken, wurden im Folgenden die flüchtigen Duftstoffe von *P. nigra* Zweigen, die entweder unbehandelt, artifiziell verletzt waren oder Eier von *D. pini* enthielten, mittels <u>GC-MS</u> analysiert. Durch Eiablage oder artifizielle Verwundung änderte sich die quantitative Zusammensetzung des Duftmusters bei *P. nigra* signifikant. Ein mit <u>multivariaten Analysemethoden</u> durchgeführter Vergleich der Duftmuster von artifiziell verwundeten und „eibelegten" *P. nigra* Zweigen und entsprechenden Zweigen von *P. sylvestris* ergab signifikante qualitative und quantitative Unterschiede. Die fehlende Reaktion von *C. ruforum* auf Düfte „eibelegter" Zweige von *P. nigra* könnte durch verschiedene Faktoren begründet sein: **(1) die falsche quantitative Zusammensetzung des Duftmusters, (2) dem Fehlen von attraktiven Schlüsselkomponenten im Duft von eierbelegten *P. nigra* oder (3) durch sog. „mixture suppression" hervorgerufen durch zusätzlich abgegebene Verbindungen, die die Attraktivität von Schlüsselkomponenten im Duftmuster maskieren** (*Kapitel 6*).

Abschließend werden die hier dargestellten eiablage-induzierten indirekten Verteidigungsmechanismen von *P. sylvestris* in Bezug zu generellen chemischen Verteidigungsmechanismen bei Kiefern gesetzt (*Kapitel 7*). Kiefern produzieren konstitutiv erhebliche Mengen an viskosem Harz, der bei einer Verwundung zu dem verletzten Gewebe transportiert und dort akkumuliert wird. Zusätzlich wird die Harz-Biosynthese induziert. Im Vergleich zu anderen Koniferen Gattungen zeigen Kiefern eine geringere induzierte Reaktion. Die durch Eiablagen induzierte Reaktion bei Kiefern ist nicht stark ausgeprägt, da signifikante Unterschiede im Duftmuster lediglich für ($E$)-β-Farnesen detektiert wurde. Dennoch haben sich offensichtlich in diesem tritrophischen System sehr fein abgestimmte und spezifische chemische Interaktionen evolviert. Einerseits lernen die Eiparasitoiden spezifisch auf ($E$)-β-Farnesen zu reagieren, wenn diese Verbindung in dem „richtigen" Kontext auftritt, d.h. vor dem Hintergrund nicht-eiablage-induzierter Kieferndüfte. Andererseits reagieren Kiefern spezifisch auf die Eiablage, wahrscheinlich aufgrund von spezifischen Elicitoren in den Oviduktsekreten der Blattwespenweibchen.

## Literatur

Hilker, M, Meiners, T (2002). Induction of plant responses towards oviposition and feeding of herbivorous arthropods: a comparison. Entomol. Exp. Appl. 104:181-192.

Hilker, M, Kobs, C, Varama, M, Schrank, K (2002a). Insect egg deposition induces *Pinus* to attract egg parasitoids. J. Exp. Biol. 205:455-461.

Hilker, M, Rohfritsch, O, Meiners, T (2002b). The plant's response towards insect egg deposition. In: Hilker, M, Meiners, T (eds), Chemoecology of Insect Eggs and Egg Deposition. Blackwell Publishing, Berlin, Oxford, pp. 205-233.

Steidle, JLM, van Loon, JJA (2003). Dietary specialization and infochemical use in carnivorous arthropods: testing a concept. Entomol. Exp. Appl. 108:133-148.

# Danksagung

Allen Mitgliedern der AG Angewandte Zoologie / Ökologie der Tiere der letzten vier Jahre möchte ich herzlich dafür danken, dass ich zu Beginn und während meiner Arbeit so überaus freundlich und offen empfangen wurde und für die Hilfsbereitschaft bei zahllosen Fragen.

Bei Prof. Dr. Monika Hilker bedanke ich mich herzlich für die Möglichkeit, dieses spannende Thema bearbeiten zu dürfen, ihr stetes Interesse und dafür dass sie viel von ihrem Wissen und ihrer Erfahrung an mich weitergegeben hat.

Ein besonderer Dank gilt Prof. Dr. Johannes Steidle für die Übernahme des Zweitgutachtens sowie für die vielen kritischen und inspirierenden Diskussionen.

Bei Dr. Torsten Meiners bedanke ich mich für die vielen fachbezogenen und weniger fachbezogenen Diskussionen und die zahlreichen kleinen „Motivationsspritzen". Dr. Joachim Ruther danke ich, dass er meinen naiven und penetranten Fragen stets mit Gelassenheit begegnet ist und die vielen guten Ratschläge, die ich darauf erhalten habe.

Ute Braun möchte ich für die ständige Hilfsbereitschaft und die hervorragende Betreuung der Zuchten danken. Frank Müller danke ich sehr für gemeinsamen „Kampf" gegen viele kleine und größere Probleme. Renate Jonas danke ich insbesondere für die vielen kleinen Aufmerksamkeiten, die sich in ihrer Summe aber als unschätzbar erweisen.

Bei Brit-Andrea Walachowicz, Claudia Stein und Tassilo Tiemann möchte ich mich für die hervorragende Zusammenarbeit während ihrer Abschlussarbeiten bedanken. Es hat riesigen Spaß gemacht! Anna Brandenburg und Ivonne Siebeke sei für ihre große Einsatzbereitschaft während der vielen Biotests herzlich gedankt.

I would like to thank also Dr. Martti Varama for collecting lots of egg parasitoids in the field during the winter months in Finland! Thanks also to Dr. Antonio Martini for his great help during my excursion to Italy and for providing sawfly cocoons. I am also grateful to Dr. Jérome Rousselet for his assistance at the excursion to France.

Prof. Dr. Stefan Schulz und Dr. Robert Wegener von der TU Braunschweig danke ich für die gute Zusammenarbeit bei der chemischen Analyse der Kieferndüfte.

Meinen Eltern danke ich dafür, dass sie stets meine Entscheidungen akzeptiert und mitgetragen haben und dafür, dass ich jede Unterstützung erhalten habe, die ich benötigte.

Sylvia Walter danke ich für alles was sie getan hat aber auch für alles was sie nicht getan hat, insbesondere aber für ihr unermessliches Verständnis und die Toleranz, die ich so manches mal auf die Probe gestellt habe.

Für die finanzielle Unterstützung danke ich der Deutschen Forschungsgemeinschaft (DFG Hi 416/11-1,2).